SECTION REVIEWS
TEACHER'S EDITION

HOLT, RINEHART AND WINSTON

A Harcourt Classroom Education Company

Austin • New York • Orlando • Atlanta • San Francisco • Boston • Dallas • Toronto • London

To the Teacher

These worksheets give students the opportunity to review the concepts presented in *Modern Chemistry*. There is one worksheet for each Chapter Section plus a mixed review worksheet that covers each complete Chapter. The worksheets are designed to be used along with the textbook to help students prepare for Chapter Tests.

Section Reviews Teacher's Edition:

Cover: Portion of the periodic table superimposed on a photomicrograph of crystals of the amino-acid dimer cystine. Photo by © Mel Pollinger/Fran Heyl Associates, NYC.

Printed in the United States of America

ISBN 0-03-057356-4

7 862 05 04

Section Reviews Pupil's Edition:

Cover: Portion of the periodic table superimposed on a photomicrograph of crystals of the amino-acid dimer cystine. Photo by © Mel Pollinger/Fran Heyl Associates, NYC.

Printed in the United States of America

ISBN 0-03-057354-8

8 9 10 862 06 05 04

TABLE OF CONTENTS

1 Matter and Change

1-1 Chemistry Is a Physical Science... 1
1-2 Matter and Its Properties.......... 3
1-3 Elements.......................... 5
Chapter 1 Mixed Review................ 7

2 Measurements and Calculations

2-1 Scientific Method.............. 9
2-2 Units of Measurement........... 11
2-3 Using Scientific Measurements.. 13
Chapter 2 Mixed Review.............. 15

3 Atoms: The Building Blocks of Matter

3-1 The Atom: From Philosophical Idea to Scientific Theory........ 17
3-2 The Structure of the Atom...... 19
3-3 Counting Atoms................. 21
Chapter 3 Mixed Review 23

4 Arrangement of Electrons in Atoms

4-1 The Development of a New Atomic Model 25
4-2 The Quantum Model of the Atom..................... 27
4-3 Electron Configurations........ 29
Chapter 4 Mixed Review.............. 31

5 The Periodic Law

5-1 History of the Periodic Table.... 33
5-2 Electron Configuration and the Periodic Table 35
5-3 Electron Configuration and Periodic Properties......... 37
Chapter 5 Mixed Review.............. 39

6 Chemical Bonding

6-1 Introduction to Chemical Bonding............. 41
6-2 Covalent Bonding and Molecular Compounds 43
6-3 Ionic Bonding and Ionic Compounds.............. 45
6-4 Metallic Bonding................ 47
6-5 Molecular Geometry............. 49
Chapter 6 Mixed Review............. 51

7 Chemical Formulas and Chemical Compounds

7-1 Chemical Names and Formulas.. 53
7-2 Oxidation Numbers............. 55
7-3 Using Chemical Formulas....... 57
7-4 Determining Chemical Formulas........................ 59
Chapter 7 Mixed Review 61

8 Chemical Equations and Reactions

8-1 Describing Chemical Reactions....................... 65
8-2 Types of Chemical Reactions 67
8-3 Activity Series of the Elements .. 69
Chapter 8 Mixed Review 71

9 Stoichiometry

9-1 Introduction to Stoichiometry... 73
9-2 Ideal Stoichiometric Calculations.................... 75
9-3 Limiting Reactants and Percent Yield 77
Chapter 9 Mixed Review............. 79

10 Physical Characteristics of Gases

10-1 The Kinetic-Molecular Theory of Matter 81
10-2 Pressure....................... 83
10-3 The Gas Laws................... 85
Chapter 10 Mixed Review............ 87

11 Molecular Composition of Gases

11-1 Volume-Mass Relationships of Gases....................... 89
11-2 The Ideal Gas Law 91
11-3 Stoichiometry of Gases 93
11-4 Effusion and Diffusion 95
Chapter 11 Mixed Review 97

12 Liquids and Solids

12-1 Liquids 99
12-2 Solids 101
12-3 Changes of State 103
12-4 Water 105
Chapter 12 Mixed Review 107

13 Solutions

13-1 Types of Mixtures 109
13-2 The Solution Process 111
13-3 Concentration of Solutions 113
Chapter 13 Mixed Review 115

14 Ions in Aqueous Solutions and Colligative Properties

14-1 Compounds in Aqueous Solutions 117
14-2 Colligative Properties of Solutions 119
Chapter 14 Mixed Review 121

15 Acids and Bases

15-1 Properties of Acids and Bases .. 123
15-2 Brønsted-Lowry Acids and Bases 125
15-3 Neutralization Reactions 127
Chapter 15 Mixed Review 129

16 Acid-Base Titration and pH

16-1 Aqueous Solutions and the Concept of pH 131
16-2 Determining pH and Titrations 133
Chapter 16 Mixed Review 135

17 Reaction Energy and Reaction Kinetics

17-1 Thermochemistry 137
17-2 Driving Force of Reactions 139
17-3 The Reaction Process 141
17-4 Reaction Rate 143
Chapter 17 Mixed Review 145

18 Chemical Equilibrium

18-1 The Nature of Chemical Equilibrium 147
18-2 Shifting Equilibrium 149
18-3 Equilibria of Acids, Bases, and Salts 151
18-4 Solubility Equilibrium 153
Chapter 18 Mixed Review 155

19 Oxidation-Reduction Reactions

19-1 Oxidation and Reduction 157
19-2 Balancing Redox Equations 159
19-3 Oxidizing and Reducing Agents 161
19-4 Electrochemistry 163
Chapter 19 Mixed Review 165

20 Carbon and Hydrocarbons

20-1 Abundance and Importance of Carbon 167
20-2 Organic Compounds 169
20-3 Saturated Hydrocarbons 171
20-4 Unsaturated Hydrocarbons 173
Chapter 20 Mixed Review 175

21 Other Organic Compounds

21-1 Functional Groups and Classes of Organic Compounds 177
21-2 More Classes of Organic Compounds 179
21-3 Organic Reactions 181
21-4 Polymers 183
Chapter 21 Mixed Review 185

22 Nuclear Chemistry

22-1 The Nucleus 187
22-2 Radioactive Decay 189
22-3 Nuclear Radiation 191
22-4 Nuclear Fission and Nuclear Fusion 193
Chapter 22 Mixed Review 195

Name ______________________ Date ______________ Class ______________

CHAPTER 1 REVIEW

Matter and Change

SECTION 1-1

SHORT ANSWER **Answer the following questions in the space provided.**

1. __a__ Technological development of a chemical product often ____.

(a) lags behind basic research on the same substance
(b) involves accidental discoveries
(c) is designed to understand a practical problem
(d) is done for the sake of learning something new

2. __d__ The primary motivation behind basic research is to ____.

(a) develop new products
(b) make money
(c) understand an environmental problem
(d) gain knowledge

3. __a__ Applied research is designed to ____.

(a) solve a particular problem
(b) develop new products
(c) gain knowledge
(d) learn for the sake of learning

4. __b__ Chemistry is ____.

(a) a biological science
(b) a physical science
(c) concerned mostly with living things
(d) the study of electricity

5. Define the six major branches of chemistry.

organic chemistry—the study of carbon-containing compounds

inorganic chemistry—the study of nonorganic substances

physical chemistry—the study of properties, changes, and the relationships between energy and matter

analytical chemistry—the identification of the composition of materials

biochemistry—the study of the chemistry of living things

theoretical chemistry—the use of mathematics and computers to design and predict the properties of new compounds

Name ______________________ Date ____________ Class ____________

SECTION 1-1 continued

6. For each of the following types of chemical investigations, determine whether the investigation is basic research, applied research, or technological development.

basic research **a.** A laboratory in a major university surveys all the reactions involving bromine.

applied research **b.** A pharmaceutical company explores a disease in order to produce a better medicine.

applied research **c.** A scientist investigates the cause of the ozone hole.

technological development **d.** A pharmaceutical company discovers a more efficient method of producing a drug.

technological development **e.** A chemical company develops a new biodegradable plastic.

applied research **f.** A laboratory explores the use of ozone to inactivate bacteria in a drinking-water system.

7. Give examples of two different instruments routinely used in chemistry.

answers may include: any type of balance, any type of microscope

8. What are microstructures?

things too small to be seen with the unaided eye

9. What is a chemical?

a compound with definite composition

10. What is chemistry?

the study of the composition, properties, and interactions of matter

Name ______________________ Date ______________ Class ______________

CHAPTER 1 REVIEW

Matter and Change

SECTION 1-2

SHORT ANSWER **Answer the following questions in the space provided.**

1. Classify each of the following as a *homogeneous* or *heterogeneous* substance:

heterogeneous **a.** iron ore

homogeneous **b.** quartz

heterogeneous **c.** granite

homogeneous **d.** soft drink

heterogeneous **e.** milk

homogeneous **f.** salt

homogeneous **g.** water

homogeneous **h.** nitrogen

2. Classify each of the following as a *physical* or *chemical* change:

physical **a.** ice melting

chemical **b.** paper burning

chemical **c.** metal rusting

physical **d.** gas under pressure

physical **e.** liquid evaporating

chemical **f.** food digesting

3. Compare a physical change with a chemical change.

A chemical change involves a rearrangement of the elements in a substance to form substances with different physical properties. A physical change may change the state of a substance but will not change the composition of that substance.

4. Compare and contrast each of the following terms:

a. *mass* and *matter*

All substances are made of matter. Mass is a measure of the amount of matter.

b. *atom* and *compound*

All matter is composed of atoms, which are the smallest units of an element that retain the properties of that element. Atoms can come together to form compounds.

c. *physical property* and *chemical property*

Physical properties are unique for a particular substance and include color, density, melting point, and boiling point. Chemical properties relate to how a substance interacts with another substance.

d. *homogeneous mixture* and *heterogeneous mixture*

A homogeneous mixture has a uniform composition. A heterogeneous mixture is not uniform.

5. Draw a diagram that compares the arrangement of atoms in the solid, liquid, and gas state.

Solid

Liquid

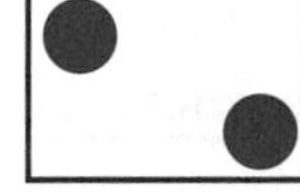

Gas

6. How is energy involved in chemical and physical changes?

Energy is either absorbed or given off in all chemical and physical changes, but it is neither created nor destroyed.

Name ______________________ Date ______________ Class ______________

CHAPTER 1 REVIEW

Matter and Change

SECTION 1-3

SHORT ANSWER **Answer the following questions in the space provided.**

1. A horizontal row of elements in the periodic table is called a(n) ___period___.

2. The symbol for the element in Period 2, Group 13, is ___B___.

3. Elements that are good conductors of heat and electricity are ___metals___.

4. Elements that are poor conductors of heat and electricity are ___nonmetals___.

5. A vertical row of elements in the periodic table is called a(n) ___group, or family___.

6. The ability of a substance to be hammered or rolled into thin sheets is called ___malleability___.

7. Would an element that is soft and able to be cut with a knife likely be a metal or a nonmetal? ___metal___.

8. Group 18 elements, which are generally unreactive, are called ___noble gases___.

9. At room temperature, most metals are ___solids___.

10. Name three characteristics of most nonmetals.

brittle, low electrical and thermal conductivity, low boiling point

11. Name three characteristics of metals.

malleable, ductile, good conductors of heat and electricity, luster

12. Name three characteristics of most metalloids.

semiconductors of electricity, solid at room temperature, less malleable than metals

13. Name two characteristics of noble gases.

generally unreactive, gas at room temperature

14. What do elements of the same group in the periodic table have in common?

Elements of the same group share similar chemical properties.

15. What do elements of the same period in the periodic table have in common?

Elements that are close to each other in the same period tend to be more similar than elements that are far apart. Physical and chemical properties change somewhat regularly across a period.

16. You are trying to manufacture a new material, but you would like to replace one of the elements in your new substance with another element that has similar chemical properties. How would you use the periodic table to choose a likely substitute?

You would consider an element of the same vertical row, or group, because elements in the same group have similar properties.

17. What is the difference between a family of elements and elements in the same period?

***Family* is another name for *group*, or elements in the same vertical row. Elements in the same period are in the same horizontal row.**

18. Complete the table below by filling in the spaces with correct names or symbols.

Name of element	Symbol of element
Aluminum	**Al**
Calcium	Ca
Manganese	Mn
Nickel	**Ni**
Phosphorus	**P**
Cobalt	**Co**
Silicon	Si
Hydrogen	H

Name ______________________ Date ______________ Class ______________

CHAPTER 1 REVIEW

Matter and Change

MIXED REVIEW

SHORT ANSWER **Answer the following questions in the space provided.**

1. Classify each of the following as a *homogeneous* or *heterogeneous* substance:

homogeneous **a.** sugar

homogeneous **b.** iron filings

heterogeneous **c.** milk

homogeneous **d.** plastic

heterogeneous **e.** cement

2. Select the most appropriate branch of chemistry from the following choices to best describe each of the investigations: organic chemistry, analytical chemistry, biochemistry, theoretical chemistry.

analytical chemistry **a.** A forensic scientist uses chemistry to find information at the scene of a crime.

theoretical chemistry **b.** A scientist uses a computer model to see how an enzyme will function.

biochemistry **c.** A professor explores the reactions that take place in a human liver.

organic chemistry **d.** An oil company scientist tries to design a better gasoline.

analytical chemistry **e.** An anthropologist tries to find out the nature of a substance in a mummy's wrap.

biochemistry **f.** A pharmaceutical company examines the protein on the coating of a virus.

3. For each of the following types of chemical investigations, determine whether the investigation is basic research, applied research, or technological development.

basic research **a.** A university plans to map all the genes on human chromosomes.

applied research **b.** A research team intends to find out why a lake remains polluted.

technological development **c.** A science teacher looks for a paint that will allow graffiti to be easily removed.

applied research **d.** A cancer research institute explores the chemistry of the cell.

basic research **e.** A professor explores the toxic compounds in marine animals.

MIXED REVIEW continued

4. Use the periodic table to identify the name, group number, and period number of the following elements:

chlorine, Group 17, Period 3 **a.** Cl

magnesium, Group 2, Period 3 **b.** Mg

tungsten, Group 6, Period 6 **c.** W

iron, Group 8, Period 4 **d.** Fe

tin, Group 14, Period 5 **e.** Sn

5. What is the difference between extensive and intensive properties?

Extensive properties depend on the amount of matter present; intensive properties do not.

6. Consider the burning of gasoline and the evaporation of gasoline. Which process represents a chemical change and which represents a physical change? Give a reason for your answer.

The burning of gasoline represents a chemical change because the gasoline is being changed into substances with different identities. Evaporation involves a physical change; the identity of gasoline remains unchanged.

7. Describe the difference between a heterogeneous mixture and a homogeneous mixture, and give an example of each.

A heterogeneous mixture, such as blood, is made of components with different physical properties. A homogeneous mixture, such as stainless steel, has a single set of physical properties.

8. Construct a concept map that includes the following terms: atom, element, compound, pure substance, mixture, homogeneous, and heterogeneous.

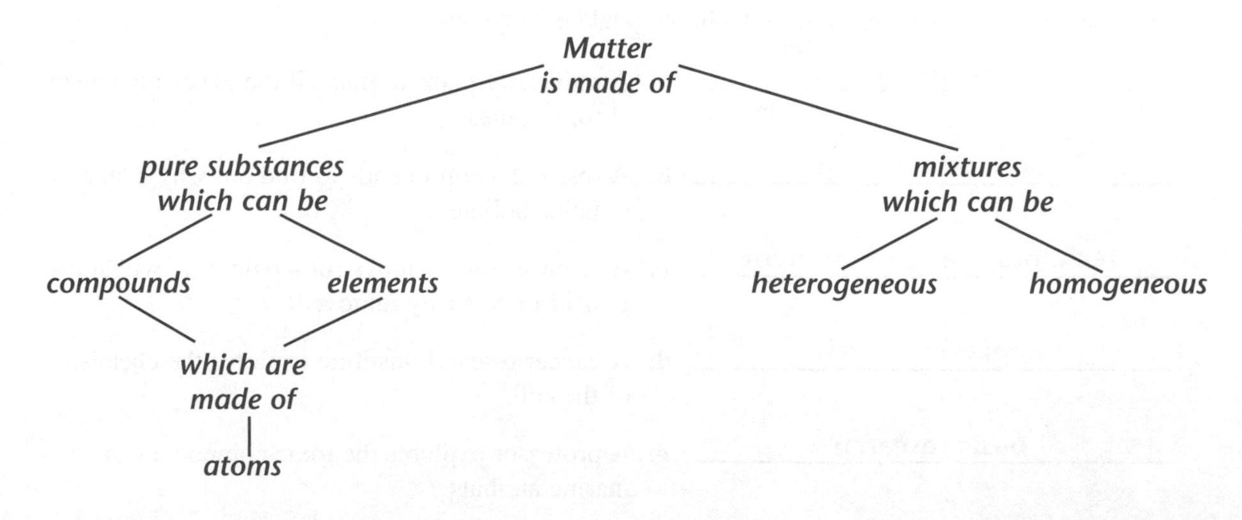

Name ______________________ Date ______________ Class ______________

0CHAPTER 2 REVIEW

Measurements and Calculations

SECTION 2-1

SHORT ANSWER **Answer the following questions in the space provided.**

1. Determine whether the following are examples of observations and data, a theory, a hypothesis, a control, or a model.

observation and data **a.** A research team records the rainfall in inches per day in a prescribed area of the rain forest. The square footage of vegetation and relative plant density per square foot are also measured.

observation and data **b.** The intensity of the precipitation, the duration, and the time of day are noted for each precipitation episode. The types of vegetation in the area are logged and classified.

control **c.** The information gathered is compared with the data on the average precipitation and the plant population collected over the last 10 years.

hypothesis **d.** The information gathered by the research team indicates that rainfall has decreased significantly. They propose that deforestation is the primary cause of this phenomenon.

2. "When 10.0 g of white, crystalline sugar is dissolved into 100. mL of water, the system is observed to freeze at −0.54°C, not 0.0°C. The system is denser than pure water." Identify which parts of these statements represent quantitative information and which parts represent qualitative information.

Quantitative values include the mass of sugar, volume of water, and observed freezing point. Qualitative properties are the color and state of the sugar and the claim of greater density.

3. Compare and contrast a model with a theory.

Theories are broad generalizations used to explain observations. Models are specific types of theories used to illustrate or explain abstract concepts.

4. Evaluate the following models. Describe how the models differ from the objects they represent.

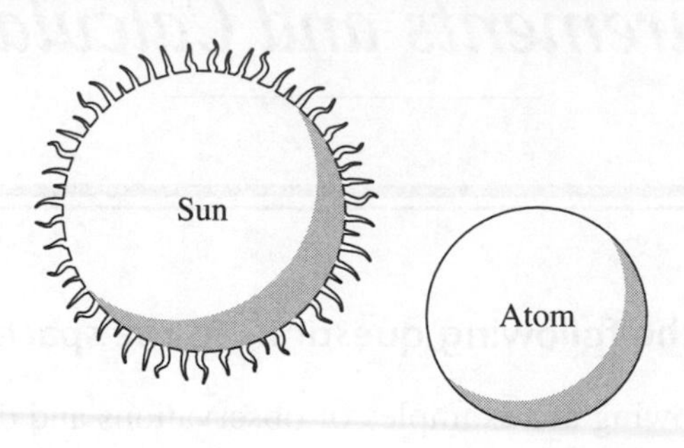

The model of the sun accurately shows that the sun is round and has a fiery surface, but the model is much smaller than the real sun and does not show the sun's composition. The model of an atom accurately shows that an atom is a particle, but the model is much larger than a real atom and does not depict an atom's composition.

5. __c__ There are ____ variables represented in the two graphs shown below.

a. one **b.** two **c.** three **d.** four

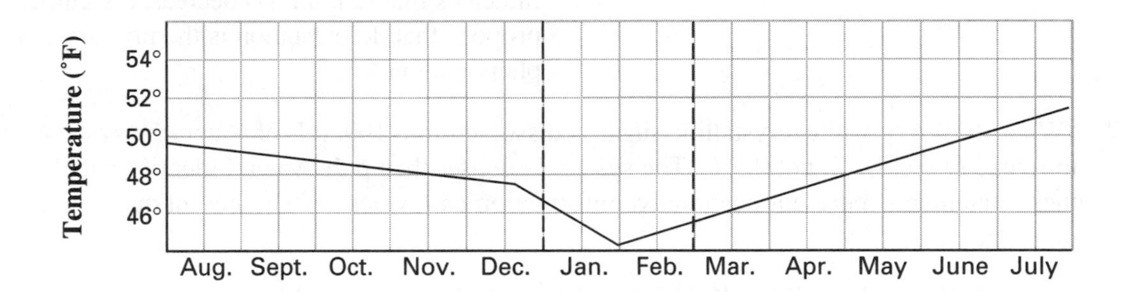

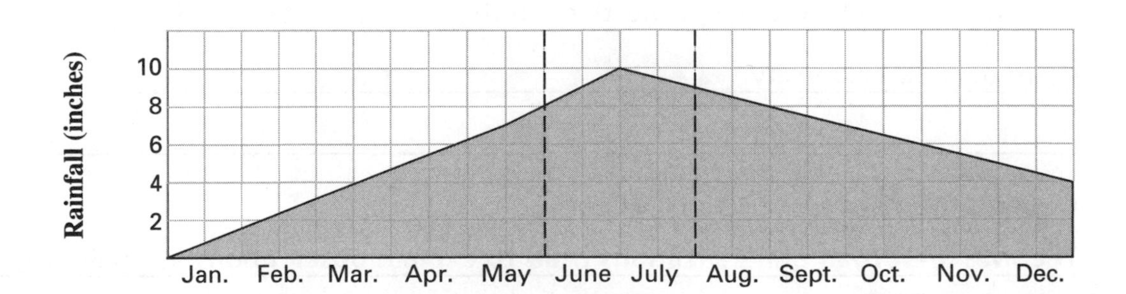

Name ______________________ Date ______________ Class ____________

CHAPTER 2 REVIEW

Measurements and Calculations

SECTION 2-2

SHORT ANSWER **Answer the following questions in the space provided.**

1. Complete the following conversions:

a. 100 mL = **0.1** L

b. 0.25 g = **25** cg

c. 400 cm^3 = **0.4** L

d. 400 cm^3 = **0.0004** m^3

2. a. Identify the quantity measured by each measuring device shown below.

A measures weight, B measures mass, and C measures volume.

b. For each quantity identified in part a, explain when it would remain constant and when it would vary.

Weight (a gravitational force) changes with altitude on Earth and when measured on a different planet or moon. In contrast, mass would be the same on the moon because gravity affects both the measured body and the mass standard equally. Volume of a liquid changes slightly with temperature and pressure.

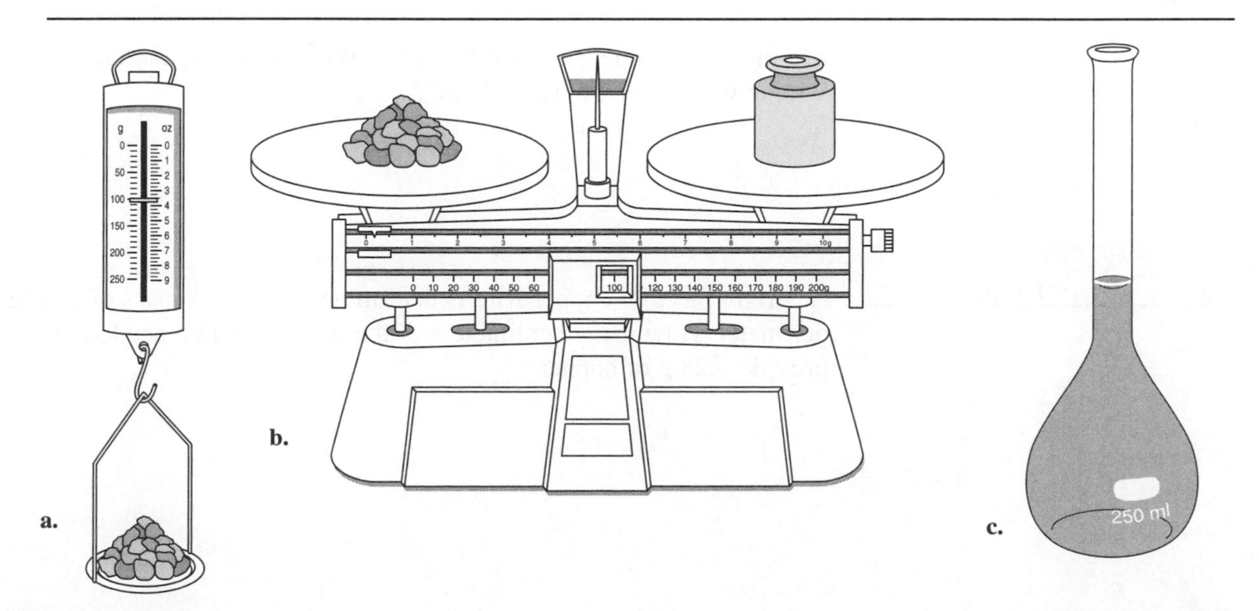

SECTION 2-2 continued

3. Use the data found in Table 2-4 on page 38 of the text to answer the following questions:

sink ______ **a.** If ice were denser than liquid water at 0°C, would it float or sink in water?

kerosene ______ **b.** Water and kerosene do not readily dissolve in one another. If the two are mixed, they quickly separate into layers. Which liquid floats on top?

gasoline, ethyl alcohol, turpentine ______ **c.** What other liquids would float on top of water?

4. Use the graph of the density of aluminum below to determine the approximate mass of aluminum samples with the following volumes.

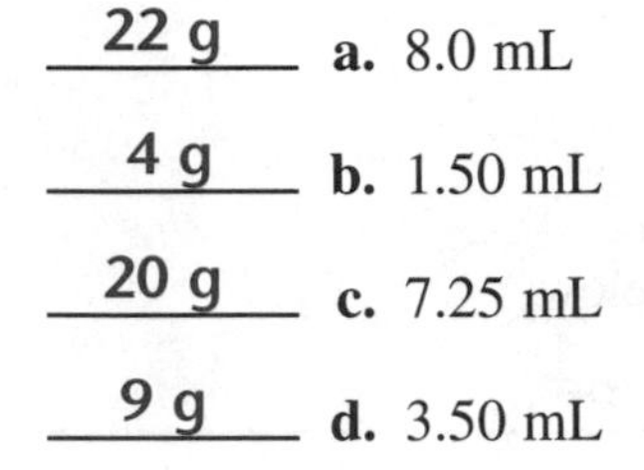

22 g ______ **a.** 8.0 mL

4 g ______ **b.** 1.50 mL

20 g ______ **c.** 7.25 mL

9 g ______ **d.** 3.50 mL

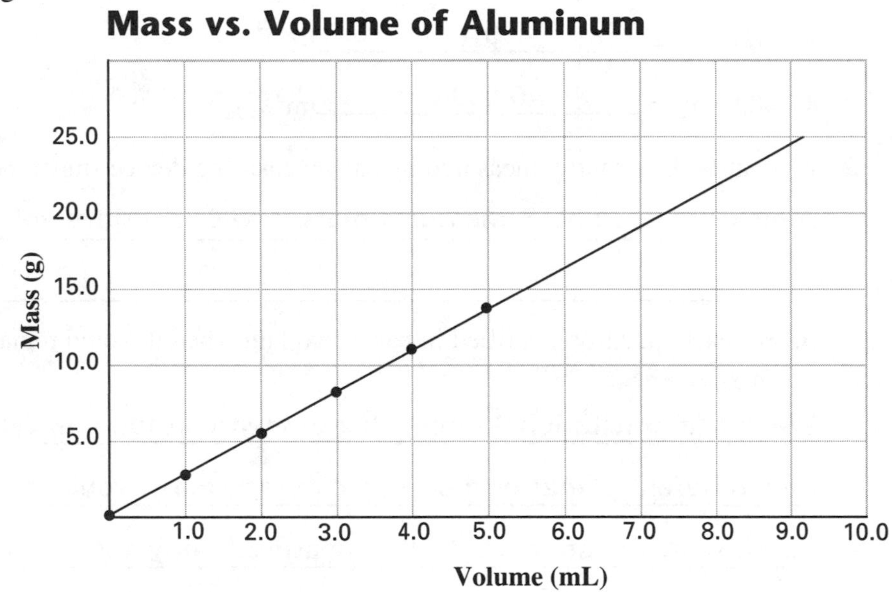

PROBLEMS **Write the answer on the line to the left. Show all your work in the space provided.**

5. 27.0 g ______ Aluminum has a density of 2.70 g/cm^3. What would be the mass of a sample whose volume is 10.0 cm^3?

6. 14 cm ______ A certain piece of copper wire is determined to have a mass of 2.00 g per meter. How many centimeters of the wire would be needed to provide 0.28 g of copper?

Name ______________________ Date ______________ Class ______________

CHAPTER 2 REVIEW

Measurements and Calculations

SECTION 2-3

SHORT ANSWER **Answer the following questions in the space provided.**

1. Report the number of significant figures in each of the following values:

<u>3</u> **a.** 0.002 37 g <u>2</u> **d.** 64 mL

<u>4</u> **b.** 0.002 037 g <u>2</u> **e.** 1.3×10^2 cm

<u>3</u> **c.** 350. J <u>3</u> **f.** 1.30×10^2 cm

2. Write the value of the following operations using scientific notation:

<u>10^{-1}</u> **a.** $\frac{10^3 \times 10^{-6}}{10^{-2}}$

<u>4×10^{-2}</u> **b.** $\frac{8 \times 10^3}{2 \times 10^5}$

<u>4.3×10^4</u> **c.** $3 \times 10^3 + 4.0 \times 10^4$

3. The following data are given for two variables, A and B:

A	B
18	2
9	4
6	6
3	12

<u>inversely proportional</u> **a.** Are A and B directly or inversely proportional?

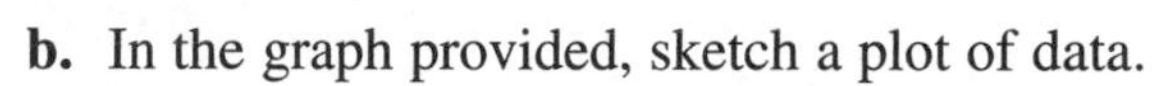

b. In the graph provided, sketch a plot of data.

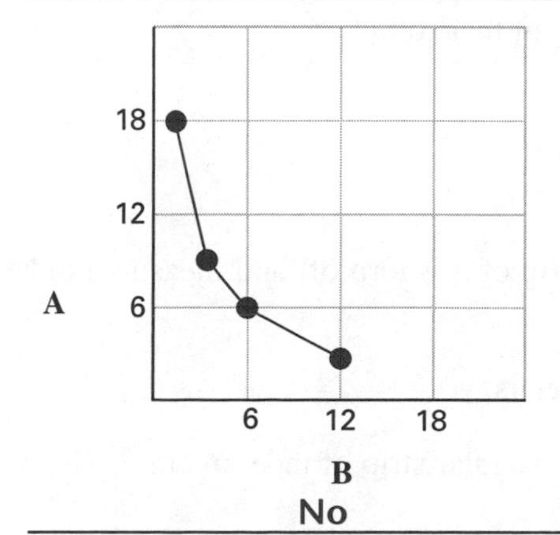

<u>No</u> **c.** Do the data points form a straight line?

<u>$A \times B = k$</u> **d.** Which equation fits the relationship shown by the data? $A \div B = k$ (a constant) or $A \times B = k$ (a constant)

<u>36</u> **e.** What is the value of k?

SECTION 2-3 continued

4. Carry out the following calculations. Express each answer to the correct number of significant figures.

____40.0 m____ **a.** 37.26 m + 2.7 m + 0.0015 m =

____2000 mL____ **b.** 256.3 mL + 2 L + 137 mL =

____151 mL____ **c.** $\frac{300.\ \text{kPa} \times 274.57\ \text{mL}}{547\ \text{kPa}} =$

____100 mL____ **d.** $\frac{346\ \text{mL} \times 200\ \text{K}}{546.4\ \text{K}} =$

5. Round the following measurements to 3 significant figures:

____22.8 g____ **a.** 22.77 g

____14.6 m____ **b.** 14.62 m

____9.31 L____ **c.** 9.3052 L

____87.6 cm____ **d.** 87.55 cm

____30.2 g____ **e.** 30.25 g

PROBLEMS Write the answer on the line to the left. Show all your work in the space provided.

6. A pure solid at a fixed temperature has a constant density. We know that

$$\text{density} = \frac{\text{mass}}{\text{volume}} \text{ or } D = \frac{m}{V}.$$

____directly proportional____ **a.** Are mass and volume directly proportional or inversely proportional for a fixed density?

____$6.0\ \text{cm}^3$____ **b.** If a solid has a density of $4.0\ \text{g/cm}^3$, what volume provides 24 g of that solid?

7. An adding-machine tape has a width of 5.80 cm. A long strip of it is torn off and measured to be 5.6 m long.

____560 cm____ **a.** Convert 5.6 m into centimeters.

____$3200\ \text{cm}^2$____ **b.** What is the area of this rectangular strip of tape, in cm^2?

Name ______________________ Date ______________ Class ______________

CHAPTER 2 REVIEW

Measurements and Calculations

MIXED REVIEW

SHORT ANSWER **Answer the following questions in the space provided.**

1. Match the description on the right to the appropriate quantity on the left.

d	2 m^3	(a)	mass of a small paper clip
a	0.5 g	(b)	length of a small paper clip
f	0.5 kg	(c)	length of a stretch limousine
e	600 cm^2	(d)	volume of a refrigerator compartment
b	20 mm	(e)	surface area of the cover of this workbook
		(f)	mass of a jar of peanut butter

2. b A measurement is said to have good precision if ____.

(a) it agrees closely with an accepted standard
(b) it agrees closely with other similar measurements
(c) it has a small number of significant figures
(d) it has a large number of significant figures

3. A certain sample with a mass of 4.00 g is found to have a volume of 7.0 mL. A student entered 4.00 ÷ 7.0 on a calculator. The calculator display shows the answer as 0.571429.

Yes **a.** Is the setup for calculating density correct?

2 **b.** How many significant figures should the answer contain?

4. It was shown in the text that in a value such as 4000 g, the precision of the number is uncertain. The zeros may or may not be significant.

1 **a.** Suppose that the mass was determined to be 4000 g to the nearest gram. How many significant figures are present in this measurement?

4.00×10^3 g **b.** Suppose next you are told that the mass in fact lies somewhere between 3950 and 4050 g. Use scientific notation to report the average value, showing an appropriate number of significant figures.

5. If you divide a sample's mass by its density, what are the resulting units?

Volume units: for example, g/(g/mL) = mL

Name ______________________ Date ______________ Class ______________

MIXED REVIEW continued

6. Three students were asked to determine the volume of a liquid by a method of their choosing. Each did three trials. The table below shows the results. (The actual volume is 24.8 mL.)

	Trial 1 (mL)	Trial 2 (mL)	Trial 3 (mL)
Student A	24.2	24.6	24.0
Student B	24.2	24.3	24.3
Student C	24.7	24.8	24.7

Student C ______ **a.** Which students' measurements showed the greatest accuracy?

Students C and B ______ **b.** Which students' measurements showed the greatest precision?

PROBLEMS Write the answer on the line to the left. Show all your work in the space provided.

7. 2.0×10^2 g ______ A single atom of platinum has a mass of 3.25×10^{-22} g. What is the mass of 6.0×10^{23} platinum atoms?

8. A sample thought to be pure lead occupies a volume of 15.0 mL and has a mass of 160.0 g.

10.7 g/mL ______ **a.** Determine its density.

No ______ **b.** Is the sample pure lead? (Refer to Table 2-4 on page 38 of the text.)

−5.7% ______ **c.** Determine the percent error, based on the accepted density for lead.

Name ________________________ Date ______________ Class ______________

CHAPTER 3 REVIEW

Atoms: The Building Blocks of Matter

SECTION 3-1

SHORT ANSWER **Answer the following questions in the space provided.**

1. Why is Democritus's view of matter considered only an idea, while Dalton's view is considered a theory?

Democritus's idea of matter does not relate atoms to a measurable property, while Dalton's theory can be tested through quantitative experimentation.

2. Give an example of a chemical or physical process that illustrates the law of conservation of mass.

A glass of ice cubes will have the same mass when the ice has completely melted into water even though its volume will change.

3. State two principles from Dalton's atomic theory that have been revised as new information has become available.

Atoms are divisible into smaller particles called subatomic particles. A given element can have atoms with different masses, called isotopes.

4. The formation of water according to the equation

$$2H_2 + O_2 \rightarrow 2H_2O$$

shows that 2 molecules (made of 4 atoms) of hydrogen and 1 molecule (made of 2 atoms) of oxygen produce 2 molecules of water. The total mass of the product, water, is equal to the sum of the masses of each of the reactants, hydrogen and oxygen. What parts of Dalton's atomic theory are illustrated by this reaction? What other law does this reaction illustrate?

Atoms cannot be subdivided, created, or destroyed. Also, atoms of different elements combine in simple, whole-number ratios to form compounds. The reaction also illustrates the law of conservation of mass.

PROBLEMS **Write the answer on the line to the left. Show all your work in the space provided.**

5. ___16 g___ If 3 g of element C combine with 8 g of element D to form compound CD, how many grams of D are needed to form compound CD_2?

6. 84.01 g of baking soda, $NaHCO_3$, *always* contains 22.99 g of sodium, 1.01 g of hydrogen, 12.01 g of carbon, and 48.00 g of oxygen. What percentage of each of these elements is present in baking soda?

___27.37%___ **a.** sodium

___1.20%___ **b.** hydrogen

___14.30%___ **c.** carbon

___57.14%___ **d.** oxygen

e. Which law do these data illustrate?

the law of definite proportions

7. Nitrogen and oxygen combine to form several compounds, as shown by the following table.

Compound	Mass of nitrogen that combines with 1 g oxygen
NO	1.7 g
NO_2	0.85 g
NO_4	0.44 g

What is the ratio of the masses of nitrogen in each of the following:

___2.0___ **a.** $\frac{NO}{NO_2}$ ___2.0___ **b.** $\frac{NO_2}{NO_4}$ ___4.0___ **c.** $\frac{NO}{NO_4}$

d. Which law do these data illustrate?

the law of multiple proportions

Name ______________________ Date ______________ Class ____________

CHAPTER 3 REVIEW

Atoms: The Building Blocks of Matter

SECTION 3-2

SHORT ANSWER **Answer the following questions in the space provided.**

1. In cathode ray tubes, the cathode ray is emitted from the negative electrode, which is called the

cathode.

2. The smallest unit of an element that can exist either alone or in combination with atoms of the same or different elements is the **atom**.

3. A positively charged particle found in the nucleus is called a(n) **proton**.

4. A nuclear particle that has no electrical charge is called a(n) **neutron**.

5. The subatomic particles that are least massive and most massive, respectively, are the

electron and **neutron**.

6. A cathode ray produced in a gas-filled tube moves away from a negative field, such as one produced by a magnet. When a paddle wheel is installed inside the tube, the wheel moves down the tube in the same direction as the cathode ray. What properties of electrons do these two phenomena illustrate?

Electrons possess charge and mass.

7. How would the electrons produced in a cathode ray tube filled with neon gas compare with the electrons produced in a cathode ray tube filled with chlorine gas?

The electrons produced from neon gas and chlorine gas would behave in the same way because electrons do not differ from element to element.

8. a. Is an atom positively charged, negatively charged, or neutral?

Atoms are neutral.

b. How does the atom maintain this charge?

Atoms consist of a positively charged nucleus, made up of protons and neutrons, that is surrounded by a negatively charged electron cloud. The positive and negative charges combine to form a net neutral charge.

Name ______________________ Date ______________ Class ______________

9. Below are two illustrations of scientists' conception of the atom. Label the electrons with a − sign and the nucleus with a + sign. On the line below the figures, identify which illustration was believed to be correct before Rutherford's gold foil experiment and which was believed to be correct after Rutherford's gold foil experiment.

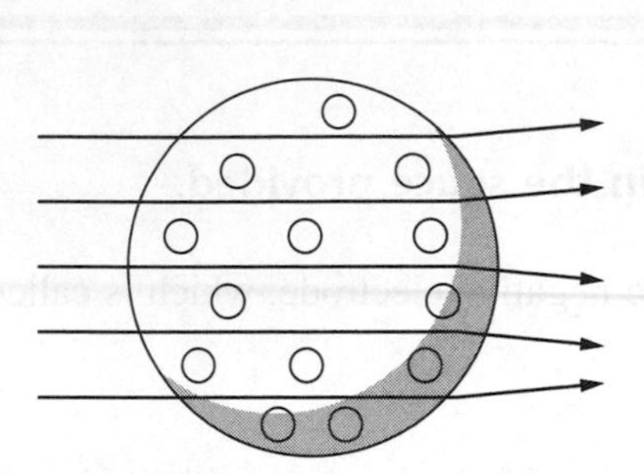

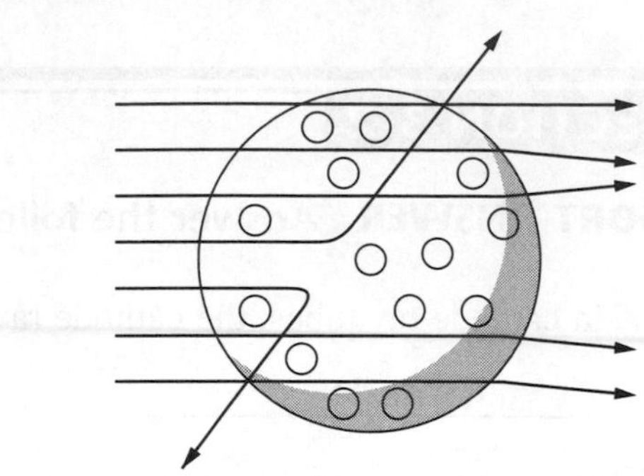

(Students should place a − sign inside all circles.)

a. **before Rutherford's experiment**

(Students should place a + sign in the center circle and a − sign in all others.)

b. **after Rutherford's experiment**

10. In the space provided, describe the locations of the subatomic particles in the labeled model of the atom below and the charge and relative mass of each particle.

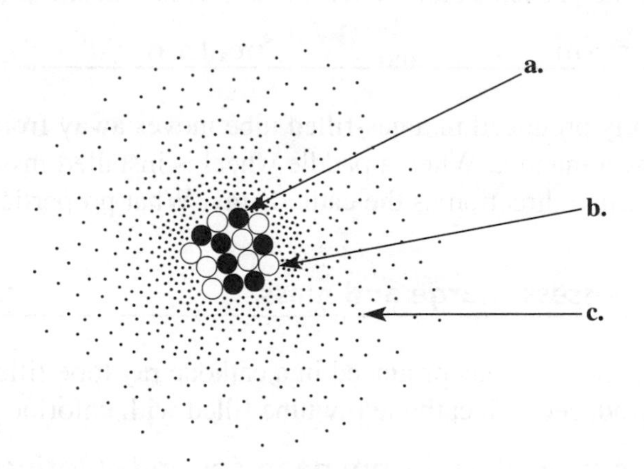

a. proton

The proton, a positive and relatively massive particle, should be located in the nucleus.

b. neutron

The neutron, a neutral and relatively massive particle, should be located in the nucleus.

c. electron

The electron, a negative particle with a low mass, should be located in the cloud surrounding the nucleus.

Name ______________________ Date ______________ Class ______________

CHAPTER 3 REVIEW

Atoms: The Building Blocks of Matter

SECTION 3-3

SHORT ANSWER **Answer the following questions in the space provided.**

1. Explain the difference between the *mass number* and the *atomic number* of a nuclide.

   Mass number is the total number of protons and neutrons in the nucleus of an isotope. Atomic number is the total number of protons only in the nucleus of each atom of an element.

2. Why is it necessary to use the average atomic mass of all isotopes rather than the mass of the most commonly occurring isotope when referring to the atomic mass of an element?

   Elements rarely occur as only one isotope; rather, they exist as mixtures of different isotopes of various masses. Using a weighted average atomic mass, you can account for the less common isotopes.

3. How many particles are in 1 mol of carbon? 1 mol of lithium? 1 mol of eggs? Will 1 mol of each of these substances have the same mass?

   There are 6.022×10^{23} particles in 1 mol of each of these substances. A mole of each substance will not have the same mass.

4. As the atomic masses of the elements in the periodic table increase, what happens to each of the following:

   a. the number of protons

   increases

   b. the number of electrons

   increases

   c. the number of atoms in 1 mol of each element

   stays the same

Name ____________________ Date ____________ Class ____________

SECTION 3-3 continued

5. Complete the following table:

Element	Symbol	Atomic number	Mass number
Europium-151	$^{151}_{63}Eu$	63	151
Silver-109	$^{109}_{47}Ag$	47	109
Tellurium-128	$^{128}_{52}Te$	52	128

6. List the number of protons, neutrons, and electrons found in zinc-66.

30 protons

36 neutrons

30 electrons

PROBLEMS **Write the answer on the line to the left. Show all your work in the space provided.**

7. 32.00 g What is the mass in grams of 2.000 mol of oxygen atoms?

8. 3.706 mol How many moles of aluminum exist in 100.0 g of aluminum?

9. 1.994×10^{24} atoms How many atoms are in 80.45 g of magnesium?

10. 1.993×10^{-21} g What is the mass in grams of 100 atoms of the carbon-12 isotope?

Name ____________________ Date ____________ Class ____________

CHAPTER 3 REVIEW

Atoms: The Building Blocks of Matter

MIXED REVIEW

SHORT ANSWER **Answer the following questions in the space provided.**

1. The element boron, B, has an atomic mass of 10.81 amu according to the periodic table. However, no single atom of boron has a mass of exactly 10.81 amu. How can you explain this difference?

The periodic table reports the average atomic mass, which is a weighted average of all isotopes of boron.

2. How did the outcome of Rutherford's gold foil experiment indicate the existence of a nucleus?

The particles rebounded, and therefore must have hit a dense bundle of matter. Because such a small percentage of particles were redirected he reasoned that this clump of matter, called the nucleus, must occupy only a small fraction of the atom's total space.

3. The ibuprofen, $C_{13}H_{18}O_2$, that is manufactured in Michigan contains 75.69% carbon, 8.80% hydrogen, and 15.51% oxygen. If you buy some ibuprofen for a headache while you are on vacation in Germany, how do you know that the ibuprofen you buy at a pharmacy overseas has the same percentage composition as the one you buy at home?

The law of definite proportions states that a chemical compound contains the same elements in exactly the same proportions by mass regardless of the site of the sample or the source of the compound.

4. Complete the following chart using the atomic mass values from the periodic table:

Compound	Mass of Fe (g)	Mass of O (g)	Ratio of O:Fe
FeO	55.85	16.00	0.2865
Fe_2O_3	111.70	48.00	0.4297
Fe_3O_4	167.55	64.00	0.3820

MIXED REVIEW continued

5. Complete the following table:

Element	Symbol	Atomic number	Mass number	Number of protons	Number of neutrons	Number of electrons
Sodium	Na	11	22	11	11	11
Fluorine	F	9	19	9	10	9
Bromine	Br	35	80	35	45	35
Calcium	Ca	20	40	20	20	20
Hydrogen	H	1	1	1	0	1
Radon	Rn	86	222	86	136	86

PROBLEMS **Write the answer on the line to the left. Show all your work in the space provided.**

6. 1.51×10^{24} atoms **a.** How many atoms are there in 2.50 mol of hydrogen?

1.51×10^{24} atoms **b.** How many atoms are there in 2.50 mol of uranium?

7. 4.65 mol How many moles are present in 107 g of sodium?

8. A certain element exists as three natural isotopes as shown in the table below.

Isotope	Mass (amu)	Percent natural abundance	Mass number
1	19.99244	90.51	20
2	20.99395	0.27	21
3	21.99138	9.22	22

20.17945 amu Calculate the average atomic mass of this element.

Name ______________________ Date ______________ Class ______________

CHAPTER 4 REVIEW

Arrangement of Electrons in Atoms

SECTION 4-1

SHORT ANSWER **Answer the following questions in the space provided.**

1. How does the photoelectric effect support the particle theory of light?

In order for an electron to be ejected from a metal surface, the electron must be struck by a single photon with at least the minimum energy needed to knock the electron loose.

2. What is the difference between the ground state and the excited state of electron positions?

The ground state is the lowest energy state the electron can occupy. When the electron absorbs energy, it can move to a higher-energy level, or excited state.

3. How can an atom emit a photon?

A photon is emitted when an electron moves from the excited state to the ground state.

4. How can the energy levels of electrons be determined by measuring the light emitted from an atom?

The frequency of the light is equal to $\frac{E}{h}$. The energy of the photon represents the difference between the energy of an electron's excited state and the energy of its ground state.

5. Why does electromagnetic radiation in the ultraviolet region represent a larger energy transition than does radiation in the infrared region?

Energy is proportional to frequency and ultraviolet radiation has a higher frequency than infrared radiation. To produce ultraviolet radiation, electrons must drop from higher energy levels than electron transitions that produce infrared radiation.

SECTION 4-1 continued

6. Which of the waves shown below has the higher frequency? Explain your answer.

The wave on the right has the higher frequency. Wavelength is inversely proportional to frequency, so as the wavelength decreases, its frequency increases.

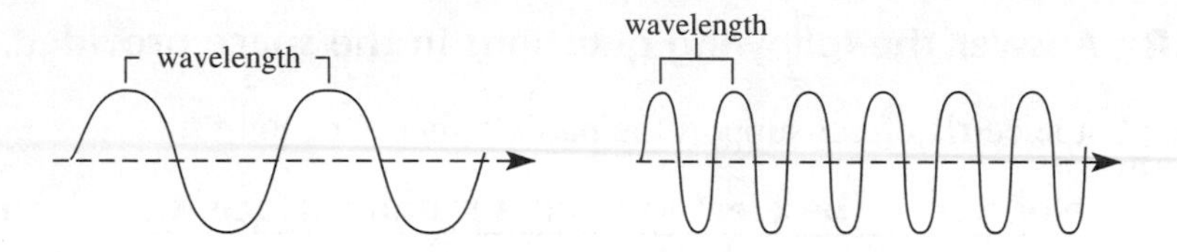

7. How many times were photons of radiation emitted from excited helium atoms to form the spectrum shown below? Explain your answer.

Photons were emitted six times. Each time an excited helium atom falls back from an excited state to its ground state or to a lower-energy state, it emits a photon of radiation that shows up as this specific line-emission spectrum. There are six lines in this helium spectrum.

Spectrum for helium

PROBLEMS Write the answer on the line to the left. Show all your work in the space provided.

8. 9.7×10^{14} Hz The wavelength of light is 310 nm. Calculate its frequency.

9. 9.4×10^{9} m What is the wavelength of electromagnetic radiation if its frequency is 3.2×10^{-2} Hz?

Name ______________________ Date ______________ Class ______________

CHAPTER 4 REVIEW

Arrangement of Electrons in Atoms

SECTION 4-2

SHORT ANSWER **Answer the following questions in the space provided.**

1. ___d___ How many quantum numbers are used to describe the energy state of an electron in an atom?

(a) 1 **(c)** 3
(b) 2 **(d)** 4

2. ___a___ A spherical electron cloud surrounding an atomic nucleus would best represent ____ .

(a) an *s* orbital **(c)** a combination of two different *p* orbitals
(b) a *p* orbital **(d)** a combination of an *s* and a *p* orbital

3. ___a___ An energy level of $n = 4$ can hold ____ electrons.

(a) 32 **(c)** 8
(b) 24 **(d)** 6

4. ___c___ An energy level of $n = 2$ can hold ____ electrons.

(a) 32 **(c)** 8
(b) 24 **(d)** 6

5. ___c___ An electron for which $n = 4$ has more ____ than an electron for which $n = 2$.

(a) spin **(c)** energy
(b) stability **(d)** wave nature

6. ___c___ According to Bohr, electrons cannot reside at ____ in the figure below.

(a) point A **(c)** point C
(b) point B **(d)** point D

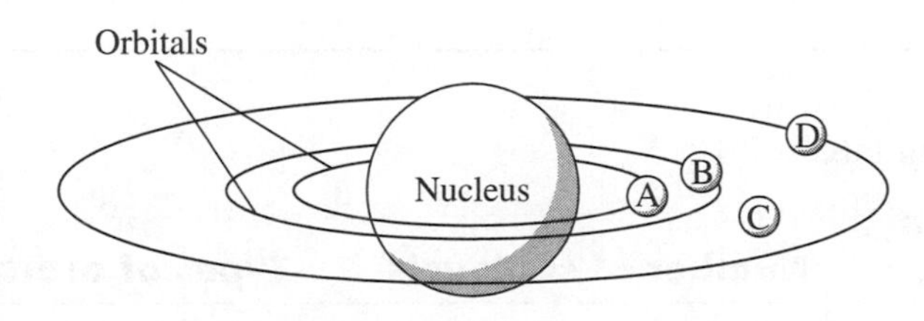

7. ___c___ According to the quantum theory, point D in the above figure represents ____.

(a) the fixed position of an electron
(b) the farthest position from the nucleus that an electron can achieve
(c) a position where an electron probably exists
(d) a position where an electron cannot exist

8. How did de Broglie conclude that electrons have a wave nature?

Scientists knew that any wave confined to a space could have only certain frequencies. De Broglie suggested that electrons should be considered as waves confined to the space around an atomic nucleus; in this way, electron waves could exist only at specific frequencies. According to the relationship $E = h\nu$, these frequencies correspond to the specific quantized energies of the Bohr orbitals.

9. Identify each of the four quantum numbers and the properties to which they refer.

The principal quantum number refers to the main energy level. The angular momentum number refers to the type of orbital the electron is in. The magnetic quantum number refers to which orbital contains the electron. The spin quantum number distinguishes between the two electrons that any orbital can hold.

10. How did the Heisenberg uncertainty principle contribute to the idea that electrons occupy "clouds," or "orbitals"?

Because the exact position of the electron is not known, it must be assumed that the electron takes up the entire space in an orbital.

11. Complete the following table.

Principal quantum number, *n*	Number of sublevels	Types of orbitals
1	1	*s*
2	2	*s,p*
3	3	*s,p,d*
4	4	*s,p,d,f*

Name ______________________ Date ______________ Class ______________

CHAPTER 4 REVIEW

Arrangement of Electrons in Atoms

SECTION 4-3

SHORT ANSWER **Answer the following questions in the space provided.**

1. Compare and contrast Hund's rule with the Pauli exclusion principle.

The Pauli exclusion principle states that no two electrons in an atom may have the same set of four quantum numbers. Hund's rule states that orbitals of equal energy are each occupied by one electron before a second electron can enter an orbital.

2. Explain the conditions under which the following orbital notation for helium is possible:

↑	↑
1*s*	2*s*

This orbital notation is possible if an electron in helium's orbital has been excited.

Write the electron configuration and orbital notation for each of the following atoms.

3. Phosphorus

$1s^22s^22p^63s^23p^3$;

↑↓	↑↓	↑↓	↑↓	↑↓	↑↓	↑	↑	↑
$1s$	$2s$	$2p_x$	$2p_y$	$2p_z$	$3s$	$3p_x$	$3p_y$	$3p_z$

4. Nitrogen

$1s^22s^22p^3$;

↑↓	↑↓	↑	↑	↑
$1s$	$2s$	$2p_x$	$2p_y$	$2p_z$

5. Potassium

$1s^22s^22p^63s^23p^64s^1$;

↑↓	↑↓	↑↓	↑↓	↑↓	↑↓	↑↓	↑↓	↑↓	↑
$1s$	$2s$	$2p_x$	$2p_y$	$2p_z$	$3s$	$3p_x$	$3p_y$	$3p_z$	$4s$

Name ______________________ Date ______________ Class ______________

SECTION 4-3 continued

6. Aluminum

$1s^2 2s^2 2p^6 3s^2 3p^1$; $\frac{\uparrow\downarrow}{1s}$ $\frac{\uparrow\downarrow}{2s}$ $\frac{\uparrow\downarrow}{2p_x}$ $\frac{\uparrow\downarrow}{2p_y}$ $\frac{\uparrow\downarrow}{2p_z}$ $\frac{\uparrow\downarrow}{3s}$ $\frac{\uparrow}{3p_x}$ $\frac{\quad}{3p_y}$ $\frac{\quad}{3p_z}$

7. Argon

$1s^2 2s^2 2p^6 3s^2 3p^6$; $\frac{\uparrow\downarrow}{1s}$ $\frac{\uparrow\downarrow}{2s}$ $\frac{\uparrow\downarrow}{2p_x}$ $\frac{\uparrow\downarrow}{2p_y}$ $\frac{\uparrow\downarrow}{2p_z}$ $\frac{\uparrow\downarrow}{3s}$ $\frac{\uparrow\downarrow}{3p_x}$ $\frac{\uparrow\downarrow}{3p_y}$ $\frac{\uparrow\downarrow}{3p_z}$

8. Boron

$1s^2 2s^2 2p^1$; $\frac{\uparrow\downarrow}{1s}$ $\frac{\uparrow\downarrow}{2s}$ $\frac{\uparrow}{2p_x}$ $\frac{\quad}{2p_y}$ $\frac{\quad}{2p_z}$

9. Which guideline, Hund's rule or the Pauli exclusion principle, is violated in the following orbital diagrams?

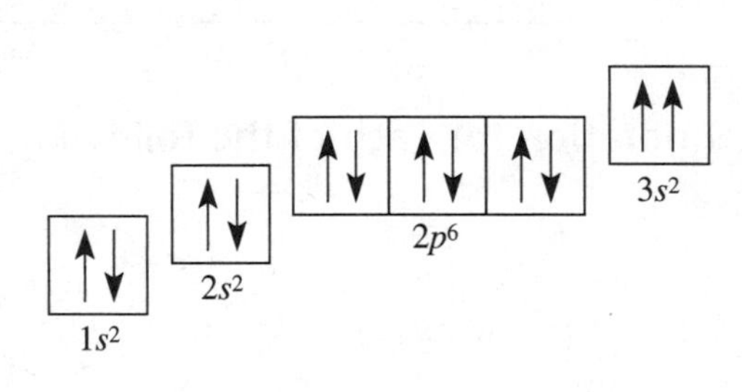

a. Pauli exclusion principle

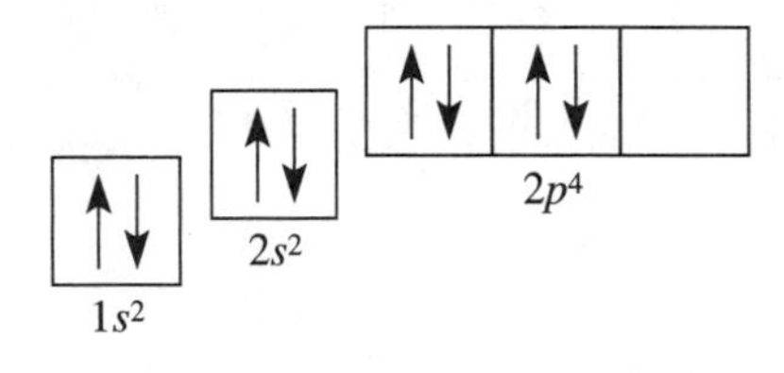

b. Hund's rule

CHAPTER 4 REVIEW

Arrangement of Electrons in Atoms

MIXED REVIEW

SHORT ANSWER **Answer the following questions in the space provided.**

1. Under what conditions does matter create light?

A photon is emitted when an electron moves from a higher-energy level to a lower-energy level.

2. What do quantum numbers describe?

Quantum numbers describe the location, type of orbital, and spin properties of electrons.

3. What is the relationship between the principal quantum number and the electron configuration?

The principle quantum number, *n*, describes the energy level. For example, the electrons at $2p^6$ are at the energy level represented by $n = 2$.

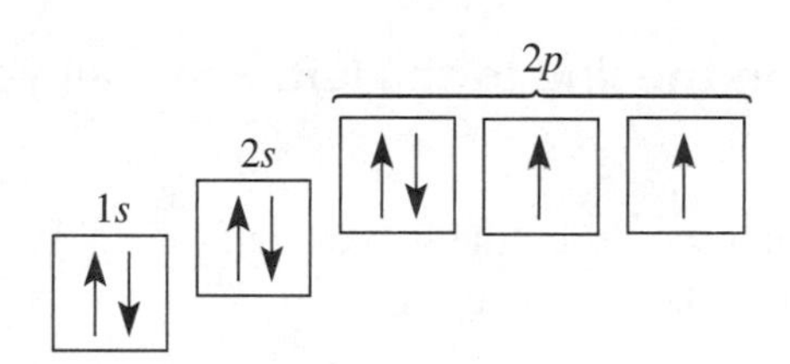

4. How does the figure above illustrate Hund's rule?

The most stable arrangement of electrons is one with the maximum number of unpaired electrons.

5. How does the figure above illustrate the Pauli exclusion principle?

No two electrons can have the same set of four quantum numbers.

Name ______________________ Date ______________ Class ______________

MIXED REVIEW continued

6. Elements of the fourth and higher main-energy levels do not seem to follow the normal sequence for filling orbitals. Why is this so?

Electrons from the *s* orbital will sometimes be promoted to a higher-energy level in order to create half-filled or filled *d* orbitals.

7. How do electrons create the colors in a line-emission spectrum?

The colors are created by photons released by an electron as the electron makes the transition from a higher-energy level to a lower-energy level.

8. Write the electron configuration of the following atoms:

a. carbon

$1s^2 2s^2 2p^2$

b. potassium

$1s^2 2s^2 2p^6 3p^6 4s^1$

c. gallium

$1s^2 2s^2 2p^6 3s^2 3p^6 4s^2 3d^{10} 4p^1$

d. copper

$1s^2 2s^2 2p^6 3s^2 3p^6 4s^1 3d^{10}$

PROBLEM **Write the answer on the line to the left. Show all your work in the space provided.**

9. 1×10^{12} m ______ What would be the wavelength of light that has a frequency of 3×10^{-4} Hz in a vacuum?

Name ______________________ Date ______________ Class ______________

CHAPTER 5 REVIEW

The Periodic Law

SECTION 5-1

SHORT ANSWER **Answer the following questions in the space provided.**

1. __c__ In the modern periodic table, elements are ordered ____.

(a) according to decreasing atomic mass
(b) according to Mendeleev's original design
(c) according to increasing atomic number
(d) based on when they were discovered

2. __d__ Mendeleev noticed that properties of elements appeared at regular intervals when the elements were arranged in order of increasing ____.

(a) density
(b) reactivity
(c) atomic number
(d) atomic mass

3. __b__ The modern periodic law states that ____.

(a) no two electrons with the same spin can be found in the same place in an atom
(b) the physical and chemical properties of the elements are functions of their atomic number
(c) electrons exhibit properties of both particles and waves
(d) the chemical properties of elements can be grouped according to periodicity, but physical properties cannot

4. __c__ The discovery of the noble gases changed Mendeleev's periodic table by adding a new ____.

(a) period
(b) series
(c) group
(d) sublevel block

5. __d__ The most distinctive property of the noble gases is that they are ____.

(a) metallic
(b) radioactive
(c) metalloids
(d) largely unreactive

6. __c__ Lithium, the first element in Group 1, has an atomic number of 3. The second element in this group has an atomic number of ____.

(a) 4
(b) 10
(c) 11
(d) 18

7. An isotope of fluorine has a mass number of 19 and an atomic number of 9.

__9__ **a.** How many protons are in this atom?

__10__ **b.** How many neutrons are in this atom?

__$^{19}_{9}F$__ **c.** What is the symbol of this fluorine atom including its mass number and atomic number?

SECTION 5-1 continued

8. Samarium, Sm, is a member of the lanthanide series.

Pu, plutonium ______ **a.** Identify the element just below samarium in the periodic table.

32 units ______ **b.** The atomic numbers of these two elements differ by how many units?

9. A certain isotope contains 53 protons, 78 neutrons, and 54 electrons.

53 ______ **a.** What is its atomic number?

127 amu ______ **b.** What is the mass of this atom in amus (to the nearest whole number)?

I ______ **c.** Is this element Pt, Xe, I, or Bh?

F, Cl, Br, At ______ **d.** Identify two other elements that are in its group.

10. In a modern periodic table, every element is a member of both a horizontal row and a vertical column. Which one is the group, and which one is the period?

The group is the vertical column, and the period is the horizontal row.

11. Explain the distinction between atomic mass and atomic number.

The atomic number is the number of protons in any atom. The atomic mass represents the mass of the protons plus the neutrons of an atom and averages this combined mass for all isotopes of that atom.

12. In the periodic table, the atomic masses of Te and I decrease rather than increase, while their atomic numbers increase. This phenomenon happens to other neighboring elements in the periodic table. Name two of these sets of elements.

Co and Ni; Ar and K; Th and Pa; U and Np; Pu and Am; Sg and Bh

Name ______________________ Date ______________ Class ______________

CHAPTER 5 REVIEW

The Periodic Law

SECTION 5-2

SHORT ANSWER **Answer the following questions in the space provided.**

Use only this periodic table to answer the following questions.

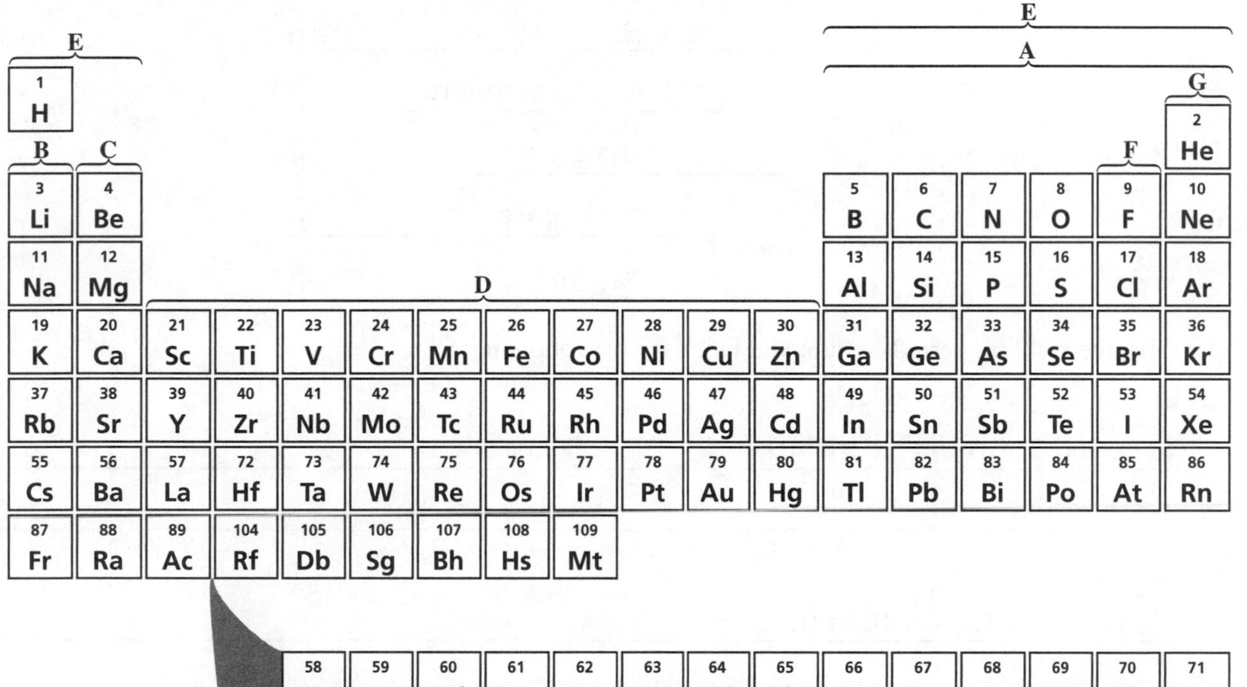

1. Identify the element, and write the noble-gas notation for each of the following:

a. the Group 14 element in Period 4

Ge; [Ar]$3d^{10}4s^{2}4p^{2}$

b. the only metal in Group 15

Bi; [Xe]$4f^{14}5d^{10}6s^{2}6p^{3}$

c. the transition metal with the smallest atomic mass

Sc; [Ar]$3d^{1}4s^{2}$

d. the alkaline earth metal with the largest atomic number

Ra; [Rn]$7s^{2}$

2. On the periodic table at the top of page 35 several areas are labeled A–H.

p block **a.** Area A represents which block, *s*, *p*, *d*, or *f*?

b. Identify the remaining labeled areas of the table, choosing from the following terms: main-group elements, transition elements, lanthanides, actinides, alkali metals, alkaline-earth metals, halogens, noble gases.

alkali metals B

alkaline-earth metals C

transition elements D

main-group elements E

halogens F

noble gases G

actinides H

3. Give the symbol, period, group, and block for the following:

a. sulfur

S, Period 3, Group 16, *p* block

b. nickel

Ni, Period 4, Group 10, *d* block

c. $[Ar]5s^1$

Rb, Period 5, Group 1, *s* block

d. $[Ne]3d^5 4s^1$

Cr, Period 4, Group 6, *d* block

4. There are 18 columns in the periodic table; each has a group number. Give the group numbers that make up each of the following blocks:

1–2 **a.** *s* block

13–18 **b.** *p* block

3–12 **c.** *d* block

Name ______________________ Date ______________ Class ______________

CHAPTER 5 REVIEW

The Periodic Law

SECTION 5-3

SHORT ANSWER **Answer the following questions in the space provided.**

1. __c__ When an electron is added to a neutral atom, energy is ____.
   (a) always absorbed
   (b) always released
   (c) either absorbed or released
   (d) burned away

2. __d__ The energy required to remove an electron from an atom is the atom's ____.
   (a) electron affinity
   (b) electron energy
   (c) electronegativity
   (d) ionization energy

3. Moving from left to right across a period on the periodic table,

   __more negative__ a. electron affinity values tend to become ____ (more negative or more positive).

   __larger__ b. ionization energy values tend to become ____ (larger or smaller).

   __smaller__ c. atomic radii tend to become ____ (larger or smaller).

4. __At__ a. Name the halogen with the least-negative electron affinity.

   __Li__ b. Name the alkali metal with the highest ionization energy.

   __Ar__ c. Name the element in Period 3 with the smallest atomic radius.

   __C__ d. Name the Group 14 element with the largest electronegativity.

5. Write the electron configuration of the following:

   a. Na

   $1s^2 2s^2 2p^6 3s^1$

   b. Na^+

   $1s^2 2s^2 2p^6$

   c. O

   $1s^2 2s^2 2p^4$

   d. O^{2-}

   $1s^2 2s^2 2p^6$

   e. Co^{2+}

   $1s^2 2s^2 2p^6 3s^2 3p^6 3d^7$

6. a. Compare the size of the radius of a positive ion to its neutral atom.

In positive ions, the radius is smaller than its corresponding neutral atom.

b. Compare the size of the radius of a negative ion to its neutral atom.

In negative ions, the radius is larger than its corresponding neutral atom.

7. a. Give the approximate positions and blocks where metals and nonmetals reside in the periodic table.

Metals are on the left-hand side of the periodic table, mostly in the *s*, *d*, and *f* blocks. Nonmetals are on the right-hand side of the periodic table, all in the *p* block.

b. Of metals and nonmetals, which tend to form positive ions? Which tend to form negative ions?

Metals tend to form positive ions; nonmetals tend to form negative ions.

8. Table 5-3 on page 145 of the text lists successive ionization energies for several elements.

$3s^2$ **a.** Identify the electron that is removed in the first ionization energy of Mg.

$3s^1$ **b.** Identify the electron that is removed in its second ionization energy.

$2p^6$ **c.** Identify the electron that is removed in its third ionization energy.

d. Explain why the second ionization energy is higher than the first, the third is higher than the second, and so on.

As electrons are removed in successive ionizations, fewer electrons remain within the atom to shield the attractive force of the nucleus. Each successive electron removed from an ion experiences an increasingly strong, effective nuclear charge.

9. Explain the role of valence electrons in the formation of chemical compounds.

Valence electrons are the electrons most subject to the influence of nearby atoms or ions. They are the electrons available to be lost, gained, or shared in the formation of chemical compounds.

Name ______________________ Date ______________ Class ______________

CHAPTER 5 REVIEW

The Periodic Law

MIXED REVIEW

SHORT ANSWER **Answer the following questions in the space provided.**

1. Consider a neutral atom with 53 protons and 74 neutrons to answer the following questions.

53 **a.** What is its atomic number?

127 amu **b.** What is its mass in amus?

atomic number **c.** Is the element's position in a modern periodic table determined by its atomic number or by its atomic mass?

2. Consider an element whose outermost electron configuration is $3d^{10}4s^24p^x$.

Period 4 **a.** To which period does the element belong?

5 **b.** If it is a halogen, what is the value of x?

True **c.** The group number will equal $(10 + 2 + x)$. True or False?

3. *p* **a.** Metalloids are found in which block, *s*, *p*, *d*, or *f*?

d **b.** The hardest, densest metals are found in which block, *s*, *p*, *d*, or *f*?

4. fluorine, F **a.** Name the most chemically active halogen.

$1s^22s^22p^5$ **b.** Write its electron configuration.

$1s^22s^22p^6$ for 1− ion **c.** Write the configuration of the most-stable ion this element makes.

5. Referring only to the periodic table at the top of the Section 5-2 Review on page 35, answer the following questions on periodic trends.

In **a.** Which has the larger radius, Al or In?

Ca **b.** Which has the larger radius, Se or Ca?

Ca **c.** Which has a larger radius, Ca or Ca^{2+}?

nonmetals **d.** Which has greater ionization energies as a class, metals or nonmetals?

Cl **e.** Which has the greater ionization energy, As or Cl?

negative ion **f.** An element with a large negative electron affinity is most likely to form a positive ion, a negative ion, or a neutral atom?

small g. In general, which has a stronger electron attraction, large atoms or small atoms?

O h. Which has greater electronegativity, O or Se?

O i. In the covalent bond between Se and O, to which atom is the electron pair more closely drawn?

6 j. How many valence electrons are there in a neutral atom of Se?

6. **Ca^{+} and Zn^{2+}** Identify all of the ions below that do not have noble gas stability.
K^{+} S^{2-} Ca^{+} I^{-} Al^{3+} Zn^{2+}

7. Using only the periodic table in Section 5-2 Review on page 35, give the noble-gas notation of the following:

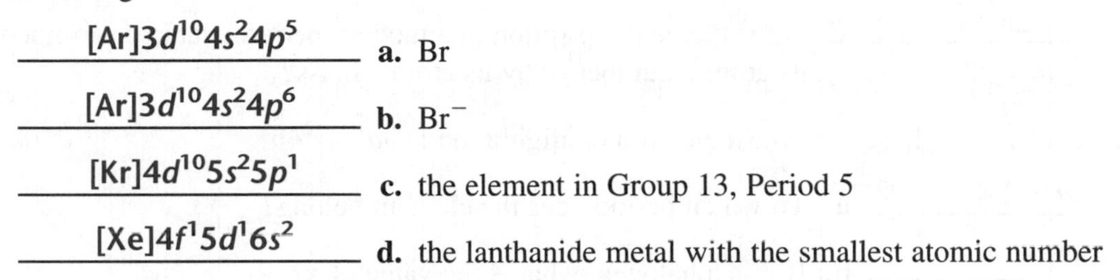

$[Ar]3d^{10}4s^{2}4p^{5}$ a. Br

$[Ar]3d^{10}4s^{2}4p^{6}$ b. Br^{-}

$[Kr]4d^{10}5s^{2}5p^{1}$ c. the element in Group 13, Period 5

$[Xe]4f^{1}5d^{1}6s^{2}$ d. the lanthanide metal with the smallest atomic number

8. Use position in the periodic table and electron configurations to describe the chemical properties of calcium and oxygen.

Calcium is an alkali metal with $[Ar]4s^{2}$ configuration. It forms a stable 2+ ion, has low ionization energy, and forms salt-like ionic compounds. Oxygen, with $[He]2s^{2}2p^{4}$ configuration, is a typical nonmetal, making a very stable 2− ion; it has high electronegativity and ionization energy and quite negative electron affinity.

9. Copper's electron configuration might be predicted to be $3d^{9}4s^{2}$. But in fact, its configuration is $3d^{10}4s^{1}$. The two elements below copper in Group 11 behave similarly. (Confirm this in the periodic table in Figure 5-6 on pages 130–131 of the text.)

$3d^{10}4s^{1}$ a. Which configuration is apparently more stable?

Yes b. Is the *d* subshell completed in the atoms of these three elements?

True c. Every element in Period 4 has four levels of electrons established. True or False?

Name ______________________ Date ______________ Class ______________

CHAPTER 6 REVIEW

Chemical Bonding

SECTION 6-1

SHORT ANSWER **Answer the following questions in the space provided.**

1. __a__ A chemical bond between atoms results from the attraction between electrons and ____.

(a) protons
(b) neutrons
(c) isotopes
(d) Lewis structures

2. __b__ A covalent bond consists of ____.

(a) a shared electron
(b) a shared electron pair
(c) two different ions
(d) an octet of electrons

3. __a__ If two covalently bonded atoms are identical, the bond is identified as ____.

(a) nonpolar covalent
(b) polar covalent
(c) nonionic
(d) dipolar

4. __b__ A covalent bond in which there is an unequal attraction for the shared electrons is ____.

(a) nonpolar
(b) polar
(c) ionic
(d) dipolar

5. __c__ Atoms with a strong attraction for electrons they share with another atom exhibit ____.

(a) zero electronegativity
(b) low electronegativity
(c) high electronegativity
(d) Lewis electronegativity

6. __c__ Bonds that possess between 5% and 50% ionic character are considered to be ____.

(a) ionic
(b) pure covalent
(c) polar covalent
(d) nonpolar covalent

7. __a__ The greater the electronegativity difference between two atoms bonded together, the greater the bond's percentage of ____.

(a) ionic character
(b) covalent character
(c) metallic character
(d) electron sharing

8. The electrons involved in the formation of a chemical bond are called __valence electrons__.

9. A chemical bond that results from the electrostatic attraction between positive and negative ions is called a(n) __ionic bond__.

Name ______________________ Date ______________ Class ______________

SECTION 6-1 continued

10. If electrons involved in bonding spend most of the time closer to one atom rather than the other, the bond is **polar covalent**.

11. If a bond's character is more than 50% ionic, then the bond is called a(n) **ionic bond**.

12. A bond's character is more than 50% ionic if the electronegativity difference between the two atoms is greater than **1.7**.

13. Write the formula for an example of each of the following compounds:

Answers will vary.

H_2 **a.** nonpolar covalent compound

HCl **b.** polar covalent compound

NaCl **c.** ionic compound

14. Describe how a covalent bond holds two atoms together.

A pair of electrons is attracted to both nuclei of the two atoms bonded together.

15. What property of the two atoms in a covalent bond determines whether or not the bond will be polar?

electronegativity

16. How can electronegativity be used to distinguish between an ionic bond and a covalent bond?

The difference between the electronegativity of the two atoms in a bond will determine whether the bond is ionic or covalent. If the difference in electronegativity is greater than 1.7, the bond is considered ionic.

17. What is unique about the bonding properties of carbon?

A carbon atom can form four covalent bonds. It also bonds with other carbon atoms to form long-chain molecules of different sizes and shapes.

CHAPTER 6 REVIEW

Chemical Bonding

SECTION 6-2

SHORT ANSWER **Answer the following questions in the space provided.**

1. Use the concept of potential energy to describe how a covalent bond forms between two atoms.

As the atoms involved in the formation of a covalent bond approach each other, the electron-proton attraction is stronger than the electron-electron and proton-proton repulsions. The atoms are drawn to each other and their potential energy is lowered. Eventually, a distance is reached at which the repulsions between the like charges equals the attraction of the opposite charges. At this point, potential energy is at a minimum and a stable molecule forms.

2. Name two elements that form compounds that are exceptions to the octet rule.

Choose from hydrogen, boron, beryllium, phosphorus, sulfur, and xenon.

3. Explain why resonance structures illustrate the limitations of Lewis structures in correctly modeling covalent bonds.

Resonance structures show that one Lewis structure cannot correctly represent the location of electrons in a bond. Resonance structures show delocalized electrons, while Lewis structures depict electrons in a definite location.

4. Bond energy is related to bond length. Use the data in the tables below to arrange the bonds listed in order of increasing bond length, from shortest bond to longest.

a.

Bond	Bond energy (kJ/mol)
H—F	569
H—I	299
H—Cl	432
H—Br	366

H—F, H—Cl, H—Br, H—I

b.

Bond	Bond energy (kJ/mol)
C—C	346
C≡C	835
C=C	612

C≡C, C=C, C—C

5. Draw Lewis structures to represent each of the following formulas:

a. NH_3

```
    ..
H—N—H
    |
    H
```

b. H_2O

```
   ..
H—O:
   |
   H
```

c. CH_4

```
   H
   ..
H:C:H
   ..
   H
```

d. C_2H_2

H:C≡C:H

e. CH_2O

```
   H
   |      ..
H—C=O
          ..
```

Name ______________________ Date ______________ Class ______________

CHAPTER 6 REVIEW

Chemical Bonding

SECTION 6-3

SHORT ANSWER **Answer the following questions in the space provided.**

1. __a__ The notation for sodium chloride, NaCl, stands for one ____.

(a) formula unit
(b) molecule
(c) crystal
(d) atom

2. __d__ In a crystal of an ionic compound, each cation is surrounded by a number of ____.

(a) molecules
(b) positive ions
(c) dipoles
(d) negative ions

3. __b__ Compared with the neutral atoms involved in the formation of an ionic compound, the crystal lattice that results is ____.

(a) higher in potential energy
(b) lower in potential energy
(c) equal in potential energy
(d) unstable

4. __b__ The lattice energy of compound A is greater than that of compound B. What can be concluded from this fact?

(a) Compound A is not an ionic compound.
(b) It will be more difficult to break the bonds in compound A than in compound B.
(c) Compound B is probably a gas.
(d) Compound A has larger crystals than compound B.

5. __b__ The forces of attraction between molecules in a molecular compound are ____.

(a) stronger than the attractive forces in ionic bonding
(b) weaker than the attractive forces in ionic bonding
(c) approximately equal to the attractive forces in ionic bonding
(d) equal to zero

6. Describe the force that holds two atoms together in an ionic bond.

The force of attraction between unlike charges holds a negative ion and a positive ion together in an ionic bond.

7. What type of energy best represents the strength of an ionic bond?

lattice energy

8. What type of bonding holds a polyatomic ion together?

The atoms in a polyatomic ion are held together with covalent bonds, but polyatomic ions combine with ions of opposite charge to form ionic compounds.

9. Arrange the ionic bonds in the table below in order of increasing strength from weakest bond to strongest.

Ionic bond	Lattice energy (kJ/mol)
NaCl	−787
CaO	−3384
KCl	−715
MgO	−3760
LiCl	−861

KCl, NaCl, LiCl, CaO, MgO

10. Draw Lewis structures for the following polyatomic ions:

a. NH_4^+

$$\left[\begin{array}{ccc} & H & \\ & | & \\ H- & N & -H \\ & | & \\ & H & \end{array}\right]^+$$

b. SO_4^{2-}

$$\left[\begin{array}{ccc} & :\ddot{O}: & \\ & | & \\ :\ddot{O}- & S & -\ddot{O}: \\ & | & \\ & :\ddot{O}: & \end{array}\right]^{2-}$$

11. Draw the two resonance structures for the nitrite anion, NO_2^-.

$$\left[\ddot{O}{=}\ddot{N}{-}\ddot{O}:\right]^- \longleftrightarrow \left[:\ddot{O}{-}\ddot{N}{=}\ddot{O}\right]^-$$

Name ______________________ Date ______________ Class ______________

CHAPTER 6 REVIEW

Chemical Bonding

SECTION 6-4

SHORT ANSWER **Answer the following questions in the space provided.**

1. __b__ In metals, the valence electrons are considered to be ____.

(a) attached to particular positive ions
(b) shared by all surrounding atoms
(c) immobile
(d) involved in covalent bonds

2. __a__ The fact that metals are malleable and ionic crystals are brittle is best explained in terms of their ____.

(a) chemical bonds
(b) London forces
(c) heats of vaporization
(d) polarity

3. __d__ As light strikes the surface of a metal, the electrons in the electron sea ____.

(a) allow the light to pass through
(b) become attached to particular positive ions
(c) fall to lower energy levels
(d) absorb and re-emit the light

4. __d__ Mobile electrons in the metallic bond are responsible for ____.

(a) luster
(b) thermal conductivity
(c) electrical conductivity
(d) all of the above

5. __a__ In general, the strength of the metallic bond ____ moving from left to right on any row of the periodic table.

(a) increases
(b) decreases
(c) remains the same

6. __c__ When a metal is drawn into a wire, the metallic bonds ____.

(a) break easily
(b) break with difficulty
(c) do not break
(d) become ionic bonds

7. Use the concept of electron configurations to explain why the number of valence electrons in metals tends to be less than the number in most nonmetals.

Most metals have their outer electrons in *s* orbitals, while nonmetals have their outer electrons in *p* orbitals.

Name ______________________ Date ______________ Class ______________

SECTION 6-4 continued

8. How does the behavior of electrons in metals contribute to the metal's ability to conduct electricity and heat?

The mobility of electrons in a network of metal atoms contributes to the metal's ability to conduct electricity and heat.

9. What is the relationship between the heat of vaporization of a metal and the strength of the bonds that hold the metal together?

The amount of heat required to vaporize a metal is a measure of the strength of the bonds that hold the metal together.

10. Draw two diagrams of a metallic bond. In the first diagram, draw a weak metallic bond; in the second, show a metallic bond that would be stronger. Be sure to include nuclear charge and number of electrons in your illustrations.

a.

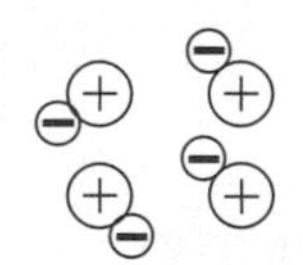

weak bond

b.

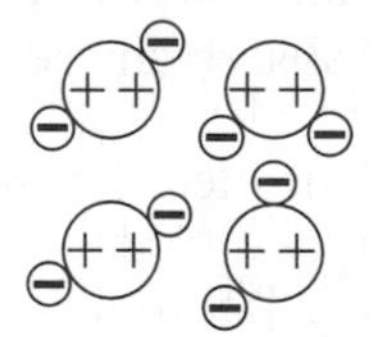

strong bond

Note: In the strong bond, the charge on the nucleus and the number of electrons must be greater than in the weak bond.

11. Complete the following table:

	Metallic	**Ionic**
Components	**atoms**	**ions**
Overall charge	**neutral**	**neutral**
Conductivity	**yes**	**no**
Melting point	**low to high**	**high**
Hardness	**soft to hard**	**hard**
Malleable	**yes**	**no**
Ductile	**yes**	**no**

Name ______________________ Date ______________ Class ______________

CHAPTER 6 REVIEW

Chemical Bonding

SECTION 6-5

SHORT ANSWER **Answer the following questions in the space provided.**

1. Identify the major assumption of the VSEPR theory that is used to predict the shape of atoms.

Pairs of electrons repel one another.

2. In water, two hydrogen atoms are bonded to one oxygen atom. Why isn't water a linear molecule?

The electron pairs that are not involved in bonding also take up space, creating a tetrahedron of electron pairs.

3. What orbitals combine together to form sp^3 hybrid orbitals around a carbon atom?

the *s* orbital and all three *p* orbitals from the second energy level

4. What two factors determine whether or not a molecule is polar?

electronegativity difference and molecular geometry

5. Arrange the following types of attractions in order of increasing strength, with 1 being the weakest and 4 the strongest.

4 covalent

3 ionic

2 dipole-dipole

1 London dispersion

6. How are dipole-dipole attractions, London dispersion forces, and hydrogen bonding similar?

In all cases there is an attraction between the slightly negative portion of one molecule and the slightly positive portion of another molecule.

Name ______________________ Date ______________ Class ______________

7. Complete the following table:

Formula	Lewis structure	Geometry	Polar
H_2S	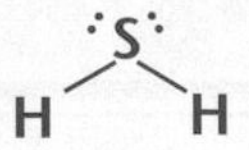 H–S–H with two lone pairs on S	bent	yes
CCl_4	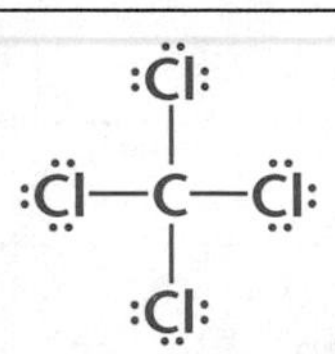 C bonded to four :Cl:	tetrahedral	no
BF_3	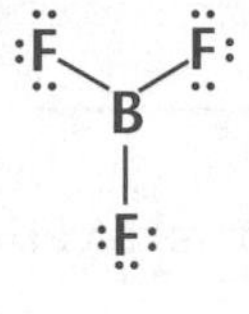 B bonded to three :F:	trigonal planar	no
H_2O	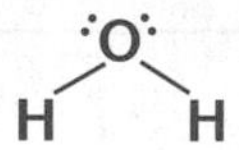 H–O–H with two lone pairs on O	bent	yes
PCl_5	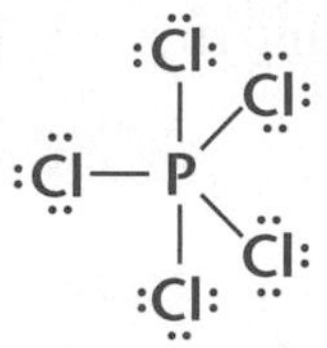 P bonded to five :Cl:	trigonal bipyramidal	no
BeF_2	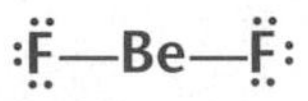 :F–Be–F:	linear	no
SF_6	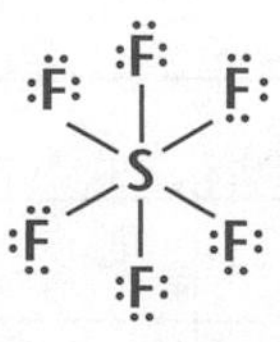 S bonded to six :F:	octahedral	no

Name ______________________ Date ______________ Class ______________

CHAPTER 6 REVIEW

Chemical Bonding

MIXED REVIEW

SHORT ANSWER **Answer the following questions in the space provided.**

1. Name the type of energy that is a measure of strength for each of the following types of bonds:

lattice energy **a.** ionic bond

bond energy **b.** covalent bond

vaporization energy **c.** metallic bond

2. Use the electronegativity values shown in Figure 5-20, on page 151 of the text, to determine whether each of the following bonds is nonpolar covalent, polar covalent, or ionic.

ionic **a.** H—F

ionic **b.** Na—Cl

polar covalent **c.** H—O

nonpolar covalent **d.** H—H

nonpolar covalent **e.** H—C

polar covalent **f.** H—N

3. How is a hydrogen bond different from an ionic or covalent bond?

A hydrogen bond is a dipole-dipole attraction between a hydrogen atom and a strongly electronegative atom such as O, N, or F. Unlike ionic or covalent bonds, in which electrons are given up completely or shared, the hydrogen bond is a weaker attraction. Hydrogen bonds are intermolecular, while ionic and covalent bonds are intramolecular.

4. H_2S and H_2O have similar structures and their central atoms belong to the same group. Yet H_2S is a gas at room temperature and H_2O is a liquid. Use bonding principles to explain why this is true.

Oxygen is more electronegative than sulfur, which creates a more-polar bond. Increased polarity in H_2O bonds means a stronger intermolecular attraction, making water a liquid at room temperature.

MIXED REVIEW continued

5. Why is a polar-covalent bond similar to an ionic bond?

There is a difference between the electronegativities of the two atoms in both types of bonds that results in electrons being more closely associated with the more electronegative atom.

6. Draw a Lewis structure for each of the following formulas. Determine whether the molecule is polar or nonpolar.

polar **a.** H_2S

:S: bonded to H and H

nonpolar **b.** $COCl_2$

:Cl: — C=O: — :Cl:

polar **c.** PCl_3

:Cl—P—Cl: with :Cl: below P

polar **d.** CH_2O

H—C=O with H above C

Name ______________________ Date ______________ Class ______________

CHAPTER 7 REVIEW

Chemical Formulas and Chemical Compounds

SECTION 7-1

SHORT ANSWER Answer the following questions in the space provided.

1. c In a Stock name such as iron(III) sulfate, the roman numeral tells us ____.

(a) how many atoms of Fe are in one formula unit
(b) how many sulfate ions can be attached to the iron atom
(c) the charge on each Fe ion
(d) the total positive charge of the formula unit

2. c The result of changing a subscript in a correctly written chemical formula is to ____.

(a) change the number of moles represented by the formula
(b) change the charges on the other ions in the compound
(c) change the formula so that it no longer represents the compound it previously represented
(d) have no effect on the formula

3. The explosive TNT has the molecular formula $C_7H_5(NO_2)_3$.

4 elements **a.** How many elements make up this compound?

6 oxygen atoms **b.** How many oxygen atoms are present in one molecule of $C_7H_5(NO_2)_3$?

21 atoms **c.** How many atoms in total are present in one molecule of $C_7H_5(NO_2)_3$?

4.2×10^{24} atoms **d.** How many atoms are present in a sample of 2×10^{23} molecules of $C_7H_5(NO_2)_3$?

4. How many atoms are present in each of these formula units?

11 atoms **a.** $Ca(HCO_3)_2$

45 atoms **b.** $C_{12}H_{22}O_{11}$

10 atoms **c.** $Fe(ClO_2)_3$

9 atoms **d.** $Fe(ClO_3)_2$

5. N_2O_5 **a.** What is the formula for the compound dinitrogen pentoxide?

carbon(IV) sulfide **b.** What is the Stock name for the covalent compound CS_2?

H_2SO_3 **c.** What is the formula for sulfurous acid?

phosphoric acid **d.** What is the name for the acid H_3PO_4?

SECTION 7-1 continued

6. Some binary compounds are ionic, others are covalent. The types of bonding partially depend on the position of the elements in the periodic table. Label each of these claims as True or False; if False, specify the nature of the error.

a. Covalently bonded binary molecular compounds typically form from nonmetals.

True

b. Binary ionic compounds form from metals and nonmetals, typically from opposite sides of the periodic table.

True

c. Binary compounds involving metalloids are always ionic.

False; metalloids form both ionic and covalent compounds.

7. Refer to Table 7-2 on page 210 of the text and Table 7-5 on page 214 of the text for examples of names and formulas for polyatomic ions and acids.

a. Derive a generalization for when an acid name will end in the suffix *-ic* or *-ous*.

In general, if the anion name ends in *-ate,* the corresponding acid name will end in a suffix of *-ic*. In general, if the anion name ends in *-ite,* the corresponding acid name will end in a suffix of *-ous*.

b. Derive a generalization for when an acid name will begin with the prefix *hydro-* and when it will not.

In general, if the anion name ends in *-ide,* the corresponding acid name will end in a suffix of *-ic* and begin with a prefix of *hydro-*. The prefix *hydro-* is never used for anions ending in *-ate* or *-ite*.

8. Fill in the blanks in the table below.

Compound name	Formula
Aluminum sulfide	**Al_2S_3**
Aluminum sulfite	**$Al_2(SO_3)_3$**
Lead(II) chloride	$PbCl_2$
Ammonium phosphate	$(NH_4)_3PO_4$
Hydroiodic acid	**HI**

Name ______________________ Date ______________ Class ______________

CHAPTER 7 REVIEW

Chemical Formulas and Chemical Compounds

SECTION 7-2

SHORT ANSWER **Answer the following questions in the space provided.**

1. Assign the oxidation number to the specified element in each of the following examples:

+4 **a.** S in H_2SO_3

+6 **b.** S in $MgSO_4$

−2 **c.** S in K_2S

+1 **d.** Cu in Cu_2S

+6 **e.** Cr in Na_2CrO_4

+1 **f.** H in $(HCO_3)^-$

+4 **g.** C in $(HCO_3)^-$

−3 **h.** N in $(NH_4)^+$

2. SCl_2 **a.** What is the formula for the compound sulfur(II) chloride?

nitrogen(IV) oxide **b.** What is the Stock name for NO_2?

3. fluorine **a.** Use electronegativity values to determine the one element that always has a negative oxidation number when it appears in any binary compound.

0; F_2 **b.** What is the oxidation number and formula for the element described in part a when it exists as an uncombined element?

4. Tin has possible oxidation numbers of +2 and +4 and forms two known oxides. One of them has the formula SnO_2.

tin(IV) oxide **a.** Give the Stock name for SnO_2.

SnO **b.** Give the empirical formula for the other oxide of tin.

5. Scientists believe that two separate reactions contribute to the depletion of the ozone layer, O_3. The first reaction involves oxides of nitrogen. The second involves free chlorine atoms. Both reactions follow. When a compound is not stated as a formula, write the correct formula in the blank beside its name.

Oxides of nitrogen:

a. NO (nitrogen monoxide) + $O_3 \rightarrow$ NO_2 (nitrogen dioxide) + O_2

Free chlorine:

b. $Cl + O_3 \rightarrow$ __ClO__ (chlorine monoxide) $+ O_2$

6. Consider the covalent compound dinitrogen trioxide when answering the following:

__N_2O_3__ **a.** What is the formula for dinitrogen trioxide?

__+3__ **b.** What is the oxidation number assigned to each N atom in this compound? Explain your answer.

The three oxygen atoms have oxidation states of −6 total, and because the algebraic sum of the oxidation states in a neutral compound must be zero, the two nitrogen atoms must have oxidation states of +6 total, therefore +3 each.

nitrogen(III) oxide **c.** Give the Stock name for dinitrogen trioxide.

7. The oxidation numbers assigned to the atoms in some organic compounds sometimes give unexpected results. Assign oxidation numbers to each atom in the following compounds:

a. CO_2

Carbon is +4, and each oxygen is −2.

b. CH_4 (methane)

Carbon is −4, and each hydrogen is +1.

c. $C_6H_{12}O_6$ (glucose)

Each carbon is 0, each hydrogen is +1, and each oxygen is −2.

d. C_3H_8 (propane gas)

Each carbon is −8/3 and each hydrogen is +1.

8. Assign oxidation numbers to each element in the compounds found in the following situations:

a. Rust, Fe_2O_3, forms on an old nail.

Each iron is +3 and each oxygen is −2.

b. Nitrogen dioxide, NO_2, pollutes the air as smog.

Nitrogen is +4 and each oxygen is −2.

c. Chromium dioxide, CrO_2, is used to make recording tapes.

Chromium is +4 and each oxygen is −2.

Name ______________________ Date ______________ Class ______________

CHAPTER 7 REVIEW

Chemical Formulas and Chemical Compounds

SECTION 7-3

SHORT ANSWER **Answer the following questions in the space provided.**

1. Label each of the following statements as True or False:

True **a.** If the formula mass of one molecule is *x* amu, the molar mass is *x* g/mol.

False **b.** Samples of two different chemicals with equal numbers of moles must have equal masses as well.

True **c.** Samples of two different chemicals with equal numbers of moles must have equal numbers of molecules as well.

2. How many moles of each element are present in a 10.0 mol sample of $Ca(NO_3)_2$?

10 mol of calcium, 20 mol of nitrogen, 60 mol of oxygen

PROBLEMS **Write the answer on the line to the left. Show all your work in the space provided.**

3. Consider a sample of 10.0 g of the gaseous hydrocarbon C_3H_4 to answer the following questions.

0.250 mol **a.** How many moles are present in this sample?

1.51×10^{23} molecules **b.** How many molecules are present in the C_3H_4 sample?

4.53×10^{23} carbon atoms **c.** How many carbon atoms are present in this sample?

10.0% **d.** What is the percentage composition of hydrogen in the sample?

4. The chief source of aluminum metal is the ore alumina, Al_2O_3.

52.9% **a.** Determine the percentage composition of Al in this ore.

2100 lb **b.** How many pounds of aluminum can be extracted from 2.0 tons of alumina?

5. Compound A has a molar mass of 20 g/mol, and compound B has a molar mass of 30 g/mol.

20 g **a.** What is the mass of 1.0 mol of compound A, in grams?

0.17 mol **b.** How many moles are present in 5.0 g of compound B?

4.0 mol **c.** How many moles of compound B are needed to have the same mass as 6.0 mol of compound A?

Name ______________ Date ______________ Class ______________

CHAPTER 7 REVIEW

Chemical Formulas and Chemical Compounds

SECTION 7-4

SHORT ANSWER **Answer the following questions in the space provided.**

1. Write empirical formulas to match the following molecular formulas:

CH_3O_2 **a.** $C_2H_6O_4$

N_2O_5 **b.** N_2O_5

HgCl **c.** Hg_2Cl_2

CH_2 **d.** C_6H_{12}

2. C_4H_8 A certain hydrocarbon has an empirical formula of CH_2 and a molar mass of 56.12 g/mol. What is its molecular formula?

3. A certain ionic compound is found to contain 0.012 mol of sodium, 0.012 mol of sulfur, and 0.018 mol of oxygen.

$Na_2S_2O_3$ **a.** What is its empirical formula?

neither **b.** Is this compound a sulfate, sulfite, or neither?

PROBLEMS **Write the answer on the line to the left. Show all your work in the space provided.**

4. Water of hydration was discussed in Sample Problem 7-11 on pages 227–228 of the text. Strong heating will drive off the water as a vapor in hydrated copper(II) sulfate. Use the data table below to answer the following:

Mass of the empty crucible	4.00 g
Mass of the crucible plus hydrate sample	4.50 g
Mass of the system after heating	4.32 g
Mass of the system after a second heating	4.32 g

36% **a.** Determine the percent water of hydration in the original sample.

5 b. The compound has the formula $CuSO_4 \cdot xH_2O$. Determine value of x.

c. What might be the purpose of the second heating?

The second heating is to ensure that all the water in the sample has been driven off.

If the mass is less after the second heating, more water was still present.

5. Gas X is found to be 24.0% carbon and 76.0% fluorine by mass.

CF_2 **a.** Determine the empirical formula of gas X.

C_4F_8 **b.** Given that the molar mass of gas X is 200.04 g/mol, determine its molecular formula.

6. A compound is found to contain 43.2% copper, 24.1% chlorine, and 32.7% oxygen by mass.

$CuClO_3$ **a.** Determine its empirical formula.

b. What is the correct Stock name of the compound in part a?

copper(I) chlorite

Name ______________________ Date ______________ Class ______________

CHAPTER 7 REVIEW

Chemical Formulas and Chemical Compounds

MIXED REVIEW

SHORT ANSWER **Answer the following questions in the space provided.**

1. Write formulas for the following compounds:

$CuCO_3$ **a.** copper(II) carbonate

Na_2SO_3 **b.** sodium sulfite

$(NH_4)_3PO_4$ **c.** ammonium phosphate

SnS_2 **d.** tin(IV) sulfide

HNO_2 **e.** nitrous acid

2. Write the Stock names for the following compounds:

magnesium perchlorate **a.** $Mg(ClO_4)_2$

iron(II) nitrate **b.** $Fe(NO_3)_2$

iron(III) nitrite **c.** $Fe(NO_2)_3$

cobalt(II) oxide **d.** CoO

nitrogen(V) oxide **e.** dinitrogen pentoxide

3. 13 atoms **a.** How many atoms are represented by the formula $Ca(HSO_4)_2$?

4.0 mol **b.** How many moles of oxygen atoms are in a 0.50 mol sample of this compound?

+6 **c.** Assign the oxidation number to sulfur in the HSO_4^- anion.

4. Assign the oxidation number to the element specified in each of the following:

+1 **a.** hydrogen in H_2O_2

−1 **b.** hydrogen in MgH_2

0 **c.** sulfur in S_8

+4 **d.** carbon in $(CO_3)^{2-}$

+6 **e.** chromium in $Na_2Cr_2O_7$

+4 **f.** nitrogen in NO_2

PROBLEMS Write the answer on the line to the left. Show all your work in the space provided.

5. c, b, d, a — Following are samples of four different compounds. Arrange them in order of increasing mass, from smallest to largest.

   a. 25 g of oxygen gas
   b. 1.00 mol of H_2O
   c. 3×10^{23} molecules of C_2H_6
   d. 2×10^{23} molecules of $C_2H_6O_2$

6. NaOH — **a.** What is the formula for sodium hydroxide?

   40.00 g/mol — **b.** What is the formula mass of sodium hydroxide?

   10. g — **c.** What is the mass in grams of 0.25 mol of sodium hydroxide?

7. 80% C, 20% H — What is the percentage composition of ethane gas, C_2H_6, to the nearest whole number?

8. $C_5H_{10}O_5$ — Ribose is an important sugar (part of RNA),with a molar mass of 150.15 g/mol. If its empirical formula is CH_2O, what is its molecular formula?

9. Butane gas, C_4H_{10}, is often used as a fuel.

174 g **a.** What is the mass in grams of 3.00 mol of butane?

1.81×10^{24} molecules **b.** How many molecules are present in that 3.00 mol sample?

C_2H_5 **c.** What is the empirical formula of the gas?

10. $C_{10}H_8$ Naphthalene is a soft covalent solid that is often used in mothballs. Its molar mass is 128.18 g/mol and it contains 93.75% carbon and 6.25% hydrogen. Determine the molecular formula of napthalene from this data.

11. Nicotine has the formula $C_xH_yN_z$. To determine its composition, a sample is burned in excess oxygen, producing the following results:

1.0 mol of CO_2
0.70 mol of H_2O
0.20 mol of NO_2

Assume that all the atoms in nicotine are present as products.

1.0 mol **a.** Determine the number of moles of carbon present in the products of this combustion.

1.40 mol ______ **b.** Determine the number of moles of hydrogen present in the combustion products.

0.20 mol ______ **c.** Determine the number of moles of nitrogen present in the combustion products.

C_5H_7N ______ **d.** Determine the empirical formula of nicotine based on your calculations.

162 g/mol ______ **e.** In a separate experiment, the molar mass of nicotine is found to be somewhere between 150 and 180 g/mol. Calculate the molar mass of nicotine to the nearest gram.

12. When $MgCO_3(s)$ is strongly heated, it produces solid MgO as gaseous CO_2 is driven off.

52.2% ______ **a.** What is the percentage loss in mass as this reaction occurs?

Mg is +2, C is +4, and O is −2 ______ **b.** Assign the oxidation number to each atom in $MgCO_3$?

No ______ **c.** Does the oxidation number of carbon change upon forming CO_2?

Name ______________________ Date ______________ Class ______________

CHAPTER 8 REVIEW

Chemical Equations and Reactions

SECTION 8-1

SHORT ANSWER **Answer the following questions in the space provided.**

1. Match the symbol on the left with its appropriate description on the right.

d	Δ	**(a)** A precipitate forms.
a	$\downarrow$	**(b)** A gas forms.
b	$\uparrow$	**(c)** A reversible reaction occurs.
f	(l)	**(d)** Heat is applied to the reactants.
e	(aq)	**(e)** A chemical is dissolved in water.
c	$\rightleftarrows$	**(f)** A chemical is in the liquid state.

2. Finish balancing the following equation:

$3Fe_3O_4$ + **8** Al $\rightarrow$ **4** Al_2O_3 + **9** Fe

3. In each of the following formulas with coefficients, write the total number of atoms present.

12 atoms **a.** $4SO_2$

16 atoms **b.** $8O_2$

51 atoms **c.** $3Al_2(SO_4)_3$

3×10^{24} atoms **d.** 6×10^{23} HNO_3

4. Convert the following word equation into a balanced chemical equation:
aluminum metal + copper(II) fluoride $\rightarrow$ aluminum fluoride + copper metal

$2Al(s) + 3CuF_2(aq) \rightarrow 2AlF_3(aq) + 3Cu(s)$

5. One way to test the salinity of a water supply is to add a few drops of silver nitrate of known concentration to the water. As the solutions of sodium chloride and silver nitrate mix, a precipitate of silver chloride forms, leaving sodium nitrate in solution. Translate these sentences into a balanced chemical equation.

$NaCl(aq) + AgNO_3(aq) \rightarrow AgCl(s) + NaNO_3(aq)$

6. a. Balance the following equation: $NaHCO_3(s) \xrightarrow{\Delta} Na_2CO_3(s) + H_2O(g) + CO_2(g)$

$2NaHCO_3(s) \xrightarrow{\Delta} Na_2CO_3(s) + H_2O(g) + CO_2(g)$

b. Translate the chemical equation in part a into a sentence.

When solid sodium hydrogen carbonate (bicarbonate) is heated, it decomposes into solid sodium carbonate while releasing carbon dioxide gas and water vapor.

7. The poisonous gas hydrogen sulfide can be neutralized with a base such as NaOH. The unbalanced equation for this reaction follows:

$$NaOH(aq) + H_2S(g) \rightarrow Na_2S(aq) + H_2O(l)$$

A student who was asked to balance this equation wrote the following:

$$Na_2OH(aq) + H_2S(g) \rightarrow Na_2S(aq) + H_3O(l)$$

Is this equation balanced? Is it correct? Explain why or why not, and supply the correct balanced equation if necessary.

It is balanced but incorrect. In two of the formulas the subscripts were changed, which changed the compounds involved. Water is not H_3O, and sodium hydroxide is not Na_2OH. The correct balanced equation is $2NaOH + H_2S \rightarrow Na_2S + 2H_2O$.

PROBLEM **Write the answer on the line to the left. Show all your work in the space provided.**

8. Recall that coefficients in a balanced equation give relative amounts of moles as well as numbers of molecules.

30 mol **a.** Calculate the amount of CO_2 in moles that forms if 10 mol of C_3H_4 react according to the following balanced equation:

$$C_3H_4 + 4O_2 \rightarrow 3CO_2 + 2H_2O$$

40 mol **b.** Calculate the amount of O_2 in moles that is consumed.

Name ______________________ Date ______________ Class ______________

CHAPTER 8 REVIEW

Chemical Equations and Reactions

SECTION 8-2

SHORT ANSWER **Answer the following questions in the space provided.**

1. Match the equation type on the left to its representation on the right.

c	synthesis	**(a)** $AX + BY \rightarrow AY + BX$
d	decomposition	**(b)** $A + BX \rightarrow AX + B$
b	single replacement	**(c)** $A + B \rightarrow AX$
a	double replacement	**(d)** $AX \rightarrow A + X$

2. c In the equation $2Al(s) + 3Fe(NO_3)_2(aq) \rightarrow 3Fe(s) + 2Al(NO_3)_3(aq)$, iron has been replaced by ____.

(a) nitrate **(c)** aluminum
(b) water **(d)** nitrogen

3. a Of the following chemical equations, the only reaction that is both synthesis and combustion is ____.

(a) $C(s) + O_2(g) \rightarrow CO_2(g)$
(b) $2C_4H_{10}(l) + 13O_2(g) \rightarrow 8CO_2(g) + 10H_2O(l)$
(c) $6CO_2(g) + 6H_2O(g) \rightarrow C_6H_{12}O_6(aq) + 6O_2(g)$
(d) $C_6H_{12}O_6(aq) + 6O_2(g) \rightarrow 6CO_2(aq) + 6H_2O(l)$

4. b Of the following chemical equations, the only reaction that is both decomposition and combustion is ____.

(a) $C(s) + O_2(g) \rightarrow CO_2(g)$
(b) $2C_4H_{10}(l) + 13O_2(g) \rightarrow 8CO_2(g) + 10H_2O(l)$
(c) $2H_2O_2(l) \rightarrow 2H_2O(l) + O_2(g)$
(d) $2HgO(s) \xrightarrow{\Delta} 2Hg(l) + O_2(g)$

5. Identify the products when the following substances decompose:

its separate elements **a.** a binary compound

metal oxide + water **b.** a metallic hydroxide

metal oxide + carbon dioxide **c.** a metallic carbonate

water + sulfur dioxide **d.** the acid H_2SO_3

6. The complete combustion of a hydrocarbon in excess oxygen yields the products CO_2 and H_2O.

SECTION 8-2 continued

7. For the following four reactions, label the type, predict the products (make sure formulas are correct), and balance the equations:

a. $Cl_2(aq) + NaI(aq) \rightarrow$

single replacement; $Cl_2(aq) + 2NaI(aq) \rightarrow I_2(aq) + 2NaCl(aq)$

b. $Mg(s) + N_2(g) \rightarrow$

synthesis; $3Mg(s) + N_2(g) \rightarrow Mg_3N_2(s)$

c. $Co(NO_3)_2(aq) + H_2S(aq) \rightarrow$

double replacement; $Co(NO_3)_2(aq) + H_2S(aq) \rightarrow CoS(s) + 2HNO_3(aq)$

d. $C_2H_5OH(aq) + O_2(g) \rightarrow$

combustion; $C_2H_5OH(aq) + 3O_2(g) \rightarrow 2CO_2(g) + 3H_2O(l)$

8. Acetylene gas, C_2H_2, is burned to provide the high temperature needed in welding.

a. Write the balanced chemical equation for the combustion of C_2H_2 in oxygen.

$2C_2H_2(g) + 5O_2(g) \rightarrow 4CO_2(g) + 2H_2O(l)$

2.0 mol **b.** If 1.0 mol of C_2H_2 is burned, how many moles of CO_2 are formed?

2.5 mol **c.** How many moles of oxygen gas are consumed?

9. Write the balanced chemical equation for the reaction that occurs when solutions of barium chloride and sodium carbonate are mixed.

$BaCl_2(aq) + Na_2CO_3(aq) \rightarrow BaCO_3(s) + 2NaCl(aq)$

10. For the commercial preparation of aluminum metal, the metal is extracted from its ore, alumina, Al_2O_3 by electrolysis. Write the balanced chemical equation for the electrolysis of molten Al_2O_3.

$2Al_2O_3(l) \rightarrow 4Al(s) + 3O_2(g)$

Name ______________________ Date ______________ Class ______________

CHAPTER 8 REVIEW

Chemical Equations and Reactions

SECTION 8-3

SHORT ANSWER **Answer the following questions in the space provided.**

1. List four metals that will *not* replace hydrogen in an acid.

Choose from Cu, Ag, Au, Pt, Sb, Bi, and Hg.

2. Consider the metals iron and silver, both listed in Table 8-3 on page 266 of the text. Which one readily forms oxides in nature, and which one does not?

Fe does form an oxide in nature, and Ag does not because it is much less active.

3. In each of the following pairs, identify the more-active element:

F_2 **a.** F_2 and I_2

K **b.** Mn and K

H **c.** Cu and H

4. Use the information in Table 8-3 on page 266 of the text to predict whether each of the following reactions will occur. For those reactions that will occur, complete the chemical equation by writing in the products formed and balancing the final equation.

a. $Al(s) + CH_3COOH(aq) \xrightarrow{50°C}$

$2Al(s) + 6CH_3COOH(aq) \xrightarrow{50°C} 2Al(CH_3COO)_3(aq) + 3H_2(g)$

b. $Al(s) + H_2O(l) \xrightarrow{50°C}$

no reaction

c. $Cr(s) + CdCl_2(aq) \rightarrow$

$2Cr(s) + 3CdCl_2(aq) \rightarrow 2CrCl_3(aq) + 3Cd(s)$

d. $Br_2(l) + KCl(aq) \rightarrow$

no reaction

5. Very active metals will react with water to release hydrogen gas.

a. Complete and then balance the equation for the reaction of Ca(*s*) with water.

$Ca(s) + 2H_2O(l) \rightarrow Ca(OH)_2(aq) + H_2(g)$

b. The reaction of rubidium, Rb, with water is faster and more violent than the reaction of Na with water. Account for this difference in terms of the two metals' atomic structure and radius.

Both are alkali metals and readily form a stable 1+ ion by ejecting an s^1 electron. Rb has a larger radius than Na and holds its electron less tightly, making it more reactive.

6. Gold is often used in jewelry. How does the relative activity of Au relate to its use in jewelry?

Gold has a low reactivity and therefore does not corrode over time.

7. Explain how to use an activity series to predict certain types of chemical behavior.

In single-displacement reactions, if the activity of the free element is greater than that of the element in the compound, the reaction will take place.

8. Aluminum is above copper in the activity series. Will aluminum metal react with copper(II) nitrate, $Cu(NO_3)_2$, to form aluminum nitrate, $Al(NO_3)_3$? If so, write the balanced equation for the reaction.

Yes; because aluminum is above copper in the activity series, aluminum metal will replace copper in copper(II) nitrate.

$2Al(s) + 3Cu(NO_3)_2(aq) \rightarrow 2Al(NO_3)_3(aq) + 3Cu(s)$

Name ______________________ Date ______________ Class ______________

CHAPTER 8 REVIEW

Chemical Equations and Reactions

MIXED REVIEW

SHORT ANSWER **Answer the following questions in the space provided.**

1. __b__ A balanced chemical equation represents all the following *except* ____.

(a) experimentally established facts
(b) the mechanism by which reactants recombine to form products
(c) identities of reactants and products in a chemical reaction
(d) relative quantities of reactants and products in a chemical reaction

2. __d__ According to the law of conservation of mass, the total mass of the reacting substance is ____.

(a) always more than the total mass of the products
(b) always less than the total mass of the products
(c) sometimes more and sometimes less than the total mass of the products
(d) always equal to the total mass of the products

3. Predict whether each of the following chemical reactions will occur. For those reactions that will occur, label the reaction type and complete the chemical equation by writing in the products formed and balancing the final equation.

a. $Ba(NO_3)_2(aq) + Na_3PO_4(aq) \rightarrow$

double replacement; $3Ba(NO_3)_2(aq) + 2Na_3PO_4(aq) \rightarrow Ba_3(PO_4)_2(s) + 6NaNO_3(aq)$

b. $Al(s) + O_2(g) \rightarrow$

synthesis; $4Al(s) + 3O_2(g) \rightarrow 2Al_2O_3(s)$

c. $I_2(s) + NaBr(aq) \rightarrow$

no reaction

d. $C_3H_4(g) + O_2(g) \rightarrow$

combustion; $C_3H_4(g) + 4O_2(g) \rightarrow 3CO_2(g) + 2H_2O(g)$

e. electrolysis of molten potassium chloride

decomposition; $2KCl(l) \rightarrow 2K(s) + Cl_2(g)$

4. Some small rockets are powered by the reaction shown in the following unbalanced equation:

$$(CH_3)_2N_2H_2(l) + N_2O_4(g) \rightarrow N_2(g) + H_2O(g) + CO_2(g) + \text{heat}$$

a. Translate this chemical equation into a sentence. (Hint: The name for $(CH_3)_2N_2H_2$ is dimethylhydrazine.)

When liquid dimethylhydrazine is mixed with dinitrogen tetroxide gas, nitrogen gas, water vapor, and gaseous carbon dioxide are produced, along with heat energy.

b. Balance the formula equation.

$(CH_3)_2N_2H_2(l) + 2N_2O_4(g) \rightarrow 3N_2(g) + 4H_2O(g) + 2CO_2(g)$

5. In the laboratory, you are given two small chips each of the unknown metals X, Y, and Z, along with dropper bottles containing solutions of $XCl_2(aq)$ and $ZCl_2(aq)$. Describe an experimental strategy you could use to determine the relative activities of X, Y, and Z.

Wording and strategies will vary. Place one chip of Y into $XCl_2(aq)$ and another into $ZCl_2(aq)$. If Y reacts with one solution but not the other, the activity series can be established. If Y replaces X but not Z, the series is $Z > Y > X$. If Y replaces Z but not X, the series is $X > Y > Z$. If Y reacts with neither solution, Y is at the bottom of the series. Then put one chip of X into $ZCl_2(aq)$. If it reacts, the series is $X > Z > Y$. If it does not react, the series is $Z > X > Y$. If Y reacts with both solutions, Y is the most reactive. Then put a chip of X into $ZCl_2(aq)$. If it reacts, the series is $Y > X > Z$. If it does not react, the series is $Y > Z > X$.

6. List the observations that would indicate that a reaction has occurred.

Signs of a reaction are evolution of heat and light, formation of a precipitate, production of a gas, etc.

Name ______________________ Date ______________ Class ______________

CHAPTER 9 REVIEW

Stoichiometry

SECTION 9-1

SHORT ANSWER **Answer the following questions in the space provided.**

1. ___b___ The coefficients in a chemical equation represent the ____.

(a) masses in grams of all reactants and products
(b) relative number of moles of reactants and products
(c) number of atoms of each element in each compound in a reaction
(d) number of valence electrons involved in a reaction

2. ___d___ Which of the following would not be studied within the topic of stoichiometry?

(a) the mole ratio of Al to Cl in the compound aluminum chloride
(b) the mass of carbon produced when a known mass of sucrose decomposes
(c) the number of moles of hydrogen that will react with a known quantity of oxygen
(d) the amount of energy required to break the ionic bonds in CaF_2

3. ___a___ A balanced chemical equation allows you to determine the ____.

(a) mole ratio of any two substances in the reaction
(b) energy released in the reaction
(c) electron configuration of all elements in the reaction
(d) reaction mechanism involved in the reaction

4. ___c___ The relative number of moles of hydrogen and oxygen that react to form water represents a(n) ____.

(a) reaction sequence
(b) bond energy
(c) mole ratio
(d) element proportion

5. Given the following unbalanced equation: $N_2O(g) + O_2(g) \rightarrow NO_2(g)$

a. Balance the equation.

$2N_2O(g) + 3O_2(g) \rightarrow 4NO_2(g)$

___4 mol NO_2:3 mol O_2___ **b.** What is the mole ratio of NO_2 to O_2?

___15.0 mol___ **c.** If 20.0 mol of NO_2 form, how many moles of O_2 must have been consumed?

___True___ **d.** Twice as many moles of NO_2 form as moles of N_2O are consumed. True or False?

___False___ **e.** Twice as many grams of NO_2 form as grams of N_2O are consumed. True or False?

SECTION 9-1 continued

PROBLEMS **Write the answer on the line to the left. Show all your work in the space provided.**

6. Given the following equation: $N_2(g) + 3H_2(g) \rightarrow 2NH_3(g)$

28.0 g/mol N_2 **a.** Determine to one decimal place the molar mass of each term, and write each one as a conversion factor from moles to grams.

2.0 g/mol H_2

17.0 g/mol NH_3

b. There are three different mole ratios in this system. Write out each one.

3 mol H_2:1 mol N_2; 2 mol NH_3:1 mol N_2; 2 mol NH_3:3 mol H_2; or their reciprocals

7. Given the following equation: $4NH_3(g) + 6NO(g) \rightarrow 5N_2(g) + 6H_2O(g)$

1 mol NO:1 mol H_2O **a.** What is the mole ratio of NO to H_2O?

3 mol NO:2 mol NH_3 **b.** What is the mole ratio of NO to NH_3?

0.360 mol **c.** If 0.240 mol of NH_3 react according to the above equation, how many moles of NO will be consumed?

8. Propyne gas can be used as a fuel. The combustion reaction of propane can be represented by the following equation:

$$C_3H_4(g) + 4O_2(g) \rightarrow 3CO_2(g) + 2H_2O(g) + \text{heat energy}$$

a. Write all the possible mole ratios in this system.

4 mol O_2:1 mol C_3H_4; 3 mol CO_2:1 mol C_3H_4; 2 mol H_2O:1 mol C_3H_4;

3 mol CO_2:4 mol O_2; 2 mol H_2O:4 mol O_2; 2 mol H_2O:3 mol CO_2;

or their reciprocals

b. Suppose that x moles of water form in the above reaction. The other three mole quantities (*not* in order) are $2x$, $1.5x$, and $0.5x$. Match these quantities to their respective formulas in the equation above.

C_3H_4 is $0.5x$; O_2 is $2x$; and CO_2 is $1.5x$

Name ______________________ Date ______________ Class ______________

CHAPTER 9 REVIEW

Stoichiometry

SECTION 9-2

PROBLEMS Write the answer on the line to the left. Show all your work in the space provided.

1. 4.5 mol The following equation represents a laboratory preparation for oxygen gas:

$2KClO_3(s) + \text{heat} \rightarrow 2KCl(s) + 3O_2(g)$

How many moles of O_2 form as 3.0 mol of $KClO_3$ are totally consumed?

2. 200 g Given the following equation: $H_2(g) + F_2(g) \rightarrow 2HF(g)$

How many grams of HF gas are produced as 5 mol of fluorine react?

3. 0.53 g Water can be made to decompose into its elements by using electricity according to the following equation:

$2H_2O(l) + \text{electrical energy} \rightarrow 2H_2(g) + O_2(g)$

How many grams of O_2 are produced as 0.033 mol of water decompose?

4. 34.8 g Sodium metal reacts with water to produce NaOH according to the following equation:

$2Na(s) + 2H_2O(l) \rightarrow 2NaOH(aq) + H_2(g)$

How many grams of NaOH are produced if 20.0 g of sodium metal react with excess oxygen?

5. ___60.2 g___ **a.** What mass of oxygen gas is produced if 100. g of lithium perchlorate are heated and allowed to decompose according to the following equation?

$$LiClO_4(s) \xrightarrow{\Delta} LiCl(s) + 2O_2(g)$$

___42.1 L___ **b.** The oxygen gas produced in part a has a density of 1.43 g/L. Calculate the volume of the O_2 gas produced.

6. A car air bag requires 70. L of nitrogen gas to inflate properly. The following equation represents the production of nitrogen gas:

$$2NaN_3(s) \rightarrow 2Na(s) + 3N_2(g)$$

___81 g___ **a.** The density of nitrogen gas is typically 1.16 g/L at room temperature. Calculate the number of grams of N_2 that are needed to inflate the air bag.

___2.9 mol___ **b.** Calculate the amount of N_2 in moles that are needed.

___130 g___ **c.** Calculate the number of grams of NaN_3 that must be used to generate the amount of nitrogen gas necessary to properly inflate the air bag.

CHAPTER 9 REVIEW

Stoichiometry

SECTION 9-3

PROBLEMS **Write the answer on the line to the left. Show all your work in the space provided.**

1. 88% If the actual yield of a reaction is 22 g and the theoretical yield is 25 g, calculate the percent yield.

2. 6.0 mol of N_2 are mixed with 12.0 mol of H_2 according to the following equation:

$$N_2(g) + 3H_2(g) \rightarrow 2NH_3(g)$$

N_2; 2.0 mol **a.** Which chemical is in excess? What is the excess amount in moles?

8.0 mol **b.** Theoretically, how many moles of NH_3 will be produced?

6.4 mol **c.** If the percent yield of NH_3 is 80%, how many moles of NH_3 are actually produced?

3. 0.050 mol of $Ca(OH)_2$ are combined with 0.080 mol of HCl according to the following equation:

$$Ca(OH)_2(aq) + 2HCl(aq) \rightarrow CaCl_2(aq) + 2H_2O(l)$$

0.10 mol **a.** How many moles of HCl are required to neutralize all 0.050 mol of $Ca(OH)_2$?

HCl b. Which is the limiting reactant in this neutralization reaction?

1.44 g c. How many grams of water will form in this reaction?

4. Acid rain can form from the combustion of nitrogen gas producing $HNO_3(aq)$ in a two-step process.

$$N_2(g) + 2O_2(g) \rightarrow 2NO_2(g)$$

$$3NO_2(g) + H_2O(g) \rightarrow 2HNO_3(aq) + NO(g)$$

1260 g a. A car burns 420. g of N_2 according to the above equations. How many grams of HNO_3 will be produced?

960 g b. For the above reactions to occur, O_2 must be in excess in the first step. What is the minimum amount of O_2 needed in grams?

690 L c. What volume does the amount of O_2 in part b occupy if its density is 1.4 g/L?

Name ______________________ Date ______________ Class ______________

CHAPTER 9 REVIEW

Stoichiometry

MIXED REVIEW

SHORT ANSWER **Answer the following questions in the space provided.**

1. Given the following equation: $C_3H_4(g) + xO_2(g) \rightarrow 3CO_2(g) + 2H_2O(g)$

4 ______ **a.** What is the value of the coefficient x in this equation?

40.07 g/mol ______ **b.** What is the molar mass of C_3H_4?

2 mol O_2:1 mol H_2O ______ **c.** What is the mole ratio of O_2 to H_2O in the above equation?

0.20 mol ______ **d.** How many moles are in an 8.0 g sample of C_3H_4?

3z ______ **e.** If z mol of C_3H_4 react, how many moles of CO_2 are produced, in terms of z?

2. a. What is meant by "ideal conditions" relative to stoichiometric calculations?

The limiting reactant is completely converted to product with no losses, as dictated by the ratio of coefficients.

b. What function do ideal stoichiometric calculations serve?

They determine the theoretical yield for the products of the reaction.

c. Are amounts actually produced typically larger or smaller than theoretical yields?

smaller

PROBLEMS **Write the answer on the line to the left. Show all your work in the space provided.**

3. Assume the reaction represented by the following equation goes all the way to completion:

$$N_2 + 3H_2 \rightarrow 2NH_3$$

4 mol ______ **a.** If 6 mol of H_2 are consumed, how many moles of NH_3 are produced?

8.5 g ______ **b.** How many grams are in a sample of NH_3 that contains 3.0×10^{23} molecules?

c. If 0.1 mol of N_2 combine with H_2, what must be true about the quantity of H_2 for N_2 to be the limiting reactant?

The moles of H_2 provided must be 0.3 or more.

4. ___**75%**___ If a reaction's theoretical yield is 8.0 g and the actual yield is 6.0 g, what is the percent yield?

5. Joseph Priestly generated oxygen gas by strongly heating mercury(II) oxide according to the following equation:

$$2HgO(s) \xrightarrow{\Delta} 2Hg(l) + O_2(g)$$

___**0.0693 mol**___ **a.** If 15.0 g HgO decompose, how many moles of HgO does this represent?

___**0.0346 mol**___ **b.** How many moles of O_2 are theoretically produced?

___**1.11 g**___ **c.** How many grams of O_2 is this?

___**0.787 L**___ **d.** If the density of O_2 gas is 1.41 g/L, how many liters of O_2 are produced?

___**1.05 g**___ **e.** If the percent yield is 95.0%, how many grams of O_2 are actually collected?

Name ________________ Date ________________ Class ________________

CHAPTER 10 REVIEW

Physical Characteristics of Gases

SECTION 10-1

SHORT ANSWER **Answer the following questions in the space provided.**

1. Identify whether the descriptions below describe an ideal gas or a real gas.

ideal gas **a.** Gas particles move in straight lines until they collide with other particles or the walls of their container.

real gas **b.** Individual gas particles have a measurable volume.

ideal gas **c.** The gas will not condense even when compressed or cooled.

ideal gas **d.** Collisions between molecules are perfectly elastic.

real gas **e.** Gas particles passing close to one another exert an attraction on each other.

2. The formula for kinetic energy is $KE = \frac{1}{2}mv^2$.

***KE* triples** **a.** What happens to the amount of kinetic energy if the mass is tripled (at constant speed)?

***KE* is reduced to $\frac{1}{4}$.** **b.** What happens to the amount of kinetic energy if the speed is halved (at constant mass)?

slower **c.** If two gases at the same temperature share the same kinetic energy, it follows that the molecules of greater mass have the ____ speed (faster or slower).

3. Use the kinetic-molecular theory to explain each of the following phenomena:

a. A strong-smelling gas released from a container in the middle of a room is soon detected in all areas of that room.

Gas molecules are in constant, rapid, random motion.

b. When 1 mol of a real gas is condensed to a liquid, the volume shrinks by a factor greater than 1000.

Molecules in a gas are far apart. They are much closer together in a liquid.

Molecules in a gas are easily squeezed closer together as the gas is compressed.

c. As a gas is heated, its rate of effusion through a small hole increases if all other factors remain constant.

As a gas is heated, each molecule's speed increases; therefore, the molecules pass through the small hole more frequently.

4. Explain why polar gas molecules experience larger deviations from ideal behavior than nonpolar molecules when all other factors (mass, temperature, etc) are held constant.

Polar molecules exert attractions on neighboring molecules and often move out of their straight-line paths because of these attractions.

5. Explain the difference in the speed-distribution curves of a gas at the two temperatures shown in the figure below.

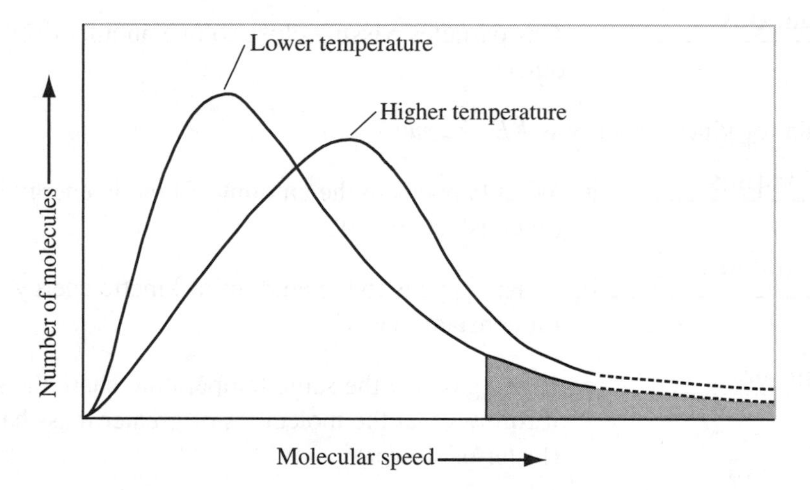

In both cases the average speed of the molecules is proportional to temperature. The distribution of molecules becomes broader as the temperature increases. This means that there are a greater number of molecules traveling within a greater range of higher speeds as the temperature increases.

Name ______________________ Date ______________ Class ______________

CHAPTER 10 REVIEW

Physical Characteristics of Gases

SECTION 10-2

SHORT ANSWER **Answer the following questions in the space provided.**

1. __b__ Pressure = force/surface area. For a constant force, when the surface area is tripled the pressure ____.

(a) is tripled
(b) is reduced by 1/3
(c) is unchanged

2. __d, c, a, b__ Rank the following pressures in increasing order:

(a) 50 kPa
(b) 2 atm
(c) 76 torr
(d) 100 N/m^2

3. Consider the following data table:

Approximate pressure (kPa)	Altitude above sea level (km)
100	0 (sea level)
50	5.5 (peak of Mt. Kilimanjaro)
25	11 (jet cruising altitude)
< 0.1	22 (ozone layer)

a. Explain briefly why the pressure decreases as the altitude increases.

As the altitude increases, there are fewer gas molecules present; therefore, there are fewer gas molecules to exert their pressure.

b. A few places on Earth are below sea level (the Dead Sea, for example). What would be true about the average atmospheric pressure there?

The pressure would exceed 100 kPa (on average) at places below sea level.

PROBLEMS **Write the answer on the line to the left. Show all your work in the space provided.**

4. Convert a pressure of 0.200 atm to the following units:

__152__ **a.** mm Hg

152 _____ **b.** torr

2.03 × 10⁴ _____ **c.** Pa

20.3 _____ **d.** kPa

5. The height of the mercury in a barometer is directly proportional to the pressure on the mercury's surface. At sea level, pressure averages 1.0 atm and the level of mercury in the barometer is 760 mm (about 30 inches). In a hurricane, the barometric reading may fall to as low as 28 inches.

0.93 atm _____ **a.** Convert a pressure reading of 28 inches to atmospheres.

380 mm Hg _____ **b.** What is the barometer reading, in mm Hg, at a pressure of 0.50 atm?

c. Can a barometer be used as an altimeter (a device for measuring altitude above sea level)? Explain your answer.

Yes; a barometer can approximate an altimeter because the higher one climbs into Earth's atmosphere, the lower the pressure recorded by the barometer.

Name ______________________ Date ______________ Class ______________

CHAPTER 10 REVIEW

Physical Characteristics of Gases

SECTION 10-3

SHORT ANSWER **Answer the following questions in the space provided.**

1. Explain how to correct for the partial pressure of water vapor when calculating the partial pressure of a dry gas that is collected over water.

Subtract the vapor pressure of water at the given collecting temperature from the atmospheric pressure taken during the collection of the gas.

2. A bicycle tire is inflated to 55 lb/in.^{2} at 15°C. Assume that the volume of the tire does not change appreciably once it is inflated.

a. If the tire and the air inside it are heated to 30°C by road wear, does the pressure in the tire increase or decrease?

The pressure increases as the temperature rises (at fixed mass and constant volume).

b. Because the temperature has doubled, does the pressure double to 110 psi?

The pressure does not double.

PROBLEMS **Write the answer on the line to the left. Show all your work in the space provided.**

3. **4.5 atm** A 24 L sample of a gas (at fixed mass and constant temperature) exerts a pressure of 3.0 atm. What pressure will the gas exert if the volume is changed to 16 L?

4. **16.7 mL** A common laboratory system to study Boyle's law uses a gas trapped in a syringe. The pressure in the system is changed by adding or removing identical weights on the plunger. The original gas volume is 50.0 mL when two weights are present. Predict the new gas volume when four more weights are added.

Name ______________________ Date ______________ Class ______________

SECTION 10-3 continued

5. a. Use 5–6 data points from Appendix Table A-8 in the text to sketch the curve for water vapor's partial pressure versus temperature on the graph provided below.

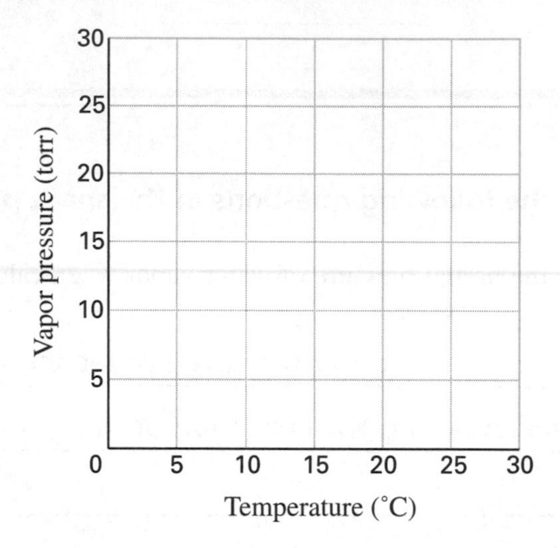

No **b.** Do the data points lie on a straight line?

10 torr **c.** Based on your sketch, predict the approximate partial pressure for water at 11°C.

6. When an explosive like TNT is detonated, a mixture of gases at high temperature is created. Suppose that gas X has a pressure of 50 atm, gas Y has a pressure of 20 atm, and gas Z has a pressure of 10 atm.

80 atm **a.** What is the total pressure in this system?

a factor of 40 **b.** Once the gas mixture combines with the air, P_{Total} soon drops to 2 atm. By what factor does the volume of the mixture increase? (Assume mass and temperature are constant.)

7. 226 mL A 250. mL sample of gas is collected at 57°C. What volume will the gas sample occupy at 25°C?

Name ______________________ Date ______________ Class ______________

CHAPTER 10 REVIEW

Physical Characteristics of Gases

MIXED REVIEW

SHORT ANSWER **Answer the following questions in the space provided.**

1. Pressure can be represented by the following equation:

$$\text{pressure} = \frac{\text{force}}{}$$

increase **a.** For a constant area, as the force increases the pressure will ____.

decrease **b.** For a constant force, as the area increases the pressure will ____.

increase **c.** For a constant pressure, as the area increases the force will ____.

2. Why are high-altitude research balloons only partially inflated before their launch?

As a balloon rises, the pressure on the outer surface will decrease. According to Boyle's law, as the pressure decreases, the volume must increase. Partially inflating the balloon allows for such expansion. It is true that the temperature decreases as it climbs, partially off-setting this expansion, but that effect is not as significant.

3. We observe that when a gas in a rigid container is warmed, the pressure on the walls increases. Explain why this occurs, based on the kinetic-molecular theory of gases.

As the temperature increases, the molecules speed up. Thus, they collide with the walls more frequently than before and with a greater force per impact. For both of these reasons, the total force per unit area increases and the pressure increases.

PROBLEMS **Write the answer on the line to the left. Show all your work in the space provided.**

4. **730 torr** A sample of chlorine gas is collected by water displacement at 23°C. If the atmospheric pressure is 751 torr, what is the partial pressure due to the chlorine?

MIXED REVIEW continued

5. ___1.80 L___ Helium gas in a balloon occupies 2.40 L at 400. K. What volume will it occupy at 300. K?

6. ___expand___ **a.** An air bubble with a volume of 2.0 mL forms at the bottom of a lake, where the pressure is 3.0 atm. As it rises, the pressure on the bubble decreases. Assume the temperature remains constant. Will the bubble expand or contract as it rises?

___6.0 mL___ **b.** Predict the volume of the bubble just as it reaches the surface, where the pressure is 1.0 atm.

7. At one point in the cycle of a piston in an automobile engine, the volume of the trapped fuel mixture is 400. cm^3 at a pressure of 1.0 atm, and a temperature of 27°C.

___9.3 atm___ **a.** In the compression of the piston, the temperature reaches 77°C and the volume decreases to 50. cm^3. What is the new pressure?

___up___ **b.** After compression, a spark plug ignites the fuel mixture, starting the power stroke of the cycle. Will the temperature of the gas go up or down?

___increase___ **c.** Will the volume of the gas increase or decrease?

___No___ **d.** Does the number of moles of gas remain constant throughout the cycle?

8. ___1.6 atm___ On a cold winter morning when the temperature is −13°C, the air pressure in an automobile tire is 1.5 atm. If the volume does not change, what will the pressure be after the tire has warmed to 13°C?

Name ______________________ Date ______________ Class ______________

CHAPTER 11 REVIEW

Molecular Composition of Gases

SECTION 11-1

SHORT ANSWER **Answer the following questions in the space provided.**

1. __d__ The volume of a gas is directly proportional to the number of moles of that gas if ____.

(a) the pressure remains constant
(b) the temperature remains constant
(c) either the pressure or temperature remains constant
(d) both the pressure and temperature remain constant

2. __c__ The molar mass of a gas at STP is the density of that gas ____.

(a) multiplied by the mass of 1 mol
(b) divided by the mass of 1 mol
(c) multiplied by 22.4 L
(d) divided by 22.4 L

3. __b__ At constant temperature and pressure, the volumes of gaseous reactants and products can be expressed ____.

(a) as the amount in moles of reactants minus the amount in moles of products
(b) as ratios of small whole numbers
(c) according to Charles's law
(d) according to Dalton's law of partial pressures

4. Two sealed flasks of equal volume, A and B, contain two different gases at the same temperature and pressure.

__True__ **a.** The two flasks must contain an equal number of molecules. True or False?

__False__ **b.** The two samples must have equal masses. True or False?

c. Now assume that flask A is warmed as flask B is cooled. Will the pressure in the two flasks remain equal? If not, which flask will have the higher pressure?

No, the pressure will not remain equal. If all other factors remain constant, flask A will have the higher pressure because it is at the higher temperature.

PROBLEMS **Write the answer on the line to the left. Show all your work in the space provided.**

5. __0.25 mol__ **a.** How many moles of methane, CH_4, are present in 5.6 L of the gas at STP?

SECTION 11-1 continued

0.25 mol **b.** How many moles of gas are present in 5.6 L of any ideal gas at STP?

4.0 g **c.** What is the mass of the 5.6 L sample of CH_4?

6. 1.96 g/L What is the density of CO_2 gas at STP?

7. 0.7 L H_2 reacts according to the following equation representing the synthesis of ammonia gas:

$$N_2(g) + 3H_2(g) \rightarrow 2NH_3(g)$$

If 1 L of H_2 is consumed, what volume of ammonia will be produced at constant temperature and pressure, based on Gay-Lussac's law of combining volumes?

8. Use the data in the table below to answer the following questions:

Formula	Molar mass (g/mol)
N_2	28.02
CO	28.01
C_2H_2	26.04
He	4.00
Ar	39.95

(Assume all gases are at STP.)

all 5 gases **a.** Which gas contains the most molecules in a 5.0 L sample?

He **b.** Which gas is the least dense?

CO and N_2 **c.** Which two gases have virtually the same density?

1.25 g/L **d.** What is that density measured at STP?

Name ____________ Date ____________ Class ____________

CHAPTER 11 REVIEW

Molecular Composition of Gases

SECTION 11-2

SHORT ANSWER **Answer the following questions in the space provided.**

1. c For the expression $V = \frac{nRT}{P}$, which of the following choices will cause the volume to increase?

(a) increase P **(c)** increase T

(b) decrease T **(d)** decrease n

2. The equation $PV = nRT$ can be rearranged to solve for any of its variables.

$P = nRT/V$ **a.** Rearrange $PV = nRT$ to solve for P.

$T = PV/nR$ **b.** Rearrange $PV = nRT$ to solve for T.

3. Explain how the ideal gas law can be simplified to give Avogadro's law, expressed as $\frac{V}{n} = k$, when the pressure and temperature of a gas are held constant.

Rearrange $PV = nRT$ to obtain $V/n = RT/P$. Because every value for RT/P is the same, its overall value is constant; therefore, $V/n = k$.

PROBLEMS **Write the answer on the line to the left. Show all your work in the space provided.**

4. 5.8 mol **a.** A large cylinder of He gas, such as that used to inflate balloons, has a volume of 25.0 L at 22°C and 5.6 atm. How many moles of He are in such a cylinder?

23 g **b.** What is the mass of the amount of helium calculated in part a?

5. ___58.2 g/mol___ **a.** Gas X has a density of 2.60 g/L at STP. Determine the molar mass of this gas.

b. Gas Y has a density of 2.60 g/L at 77°C and 0.80 atm. Its molar mass is not necessarily the same as that of gas X, even though the gases have the same density. Explain why the density is the same for X and Y but the molar masses differ.

One mole of a gas that is not at STP does not necessarily occupy 22.4 L. Multiplying the density of the gas by 22.4 L when the gas is not at STP will not give a correct molar mass.

___93 g/mol___ **c.** Determine the molar mass of gas Y.

6. ___0.25 mol___ **a.** How many moles of methane gas, CH_4, are present in a 4.0 g sample?

___1600 torr___ **b.** Find the pressure, in torr, exerted by the CH_4 gas sample when its temperature is 27°C and its volume is 3000. mL.

7. ___2.92×10^3 kPa___ A 7.00 L sample of argon gas at 420. K exerts a pressure of 625 kPa. If the gas is compressed to 1.25 L and the temperature is lowered to 350. K, what will be its new pressure?

Name ______________________ Date ______________ Class ______________

CHAPTER 11 REVIEW

Molecular Composition of Gases

SECTION 11-3

SHORT ANSWER **Answer the following questions in the space provided.**

1. ___c___ Volumes of gaseous reactants and products in a chemical reaction can be expressed as ratios of small whole numbers if ____.

(a) all reactants and products are gases
(b) standard temperature and pressure are maintained
(c) constant temperature and pressure are maintained
(d) all masses represent 1 mol quantities

2. ___a___ In the reaction $2H_2(g) + O_2(g) \rightarrow 2H_2O(g)$, the volume ratio of H_2 to H_2O is ____.

(a) 1:1
(b) 2:1
(c) 2:3
(d) 4:3
(e) 4:6

3. Ethanol is a component of gasohol, a type of fuel. The equation representing its complete combustion follows:

$$C_2H_6O(g) + xO_2(g) \rightarrow 2CO_2(g) + 3H_2O(l)$$

___3___ **a.** What is the value of x when the equation is correctly balanced?

___the volume of O_2 consumed___ **b.** Which is greater, the volume of O_2 consumed or the volume of CO_2 produced? (Assume reactants and products exist under the same conditions.)

4. The following reaction is carried out in a flexible container:

$$H_2(g) + I_2(g) \rightarrow 2HI(g)$$

The temperature and pressure are held constant throughout the reaction. Will the volume of the container increase, decrease, or remain the same as the reaction proceeds? Explain your answer.

The volume will remain the same as the reaction proceeds. The total volume of gases consumed as reactants equals the total volume of HI produced; therefore no net change occurs as long as temperature and pressure remain constant.

Name ______________________ Date ______________ Class ______________

SECTION 11-3 continued

PROBLEMS **Write the answer on the line to the left. Show all your work in the space provided.**

5. When C_3H_4 combusts at STP, 5.6 L of C_3H_4 are consumed according to the following equation:

$$C_3H_4(g) + 4O_2(g) \rightarrow 3CO_2(g) + 2H_2O(l)$$

0.25 mol **a.** How many moles of C_3H_4 react?

1.0 mol of O_2

0.75 mol of CO_2

0.50 mol of H_2O

b. How many moles of O_2, CO_2, and H_2O are either consumed or produced in the above reaction?

10. g **c.** How many grams of C_3H_4 are consumed?

17 L **d.** How many liters of CO_2 are produced?

9.0 g **e.** How many grams of H_2O are produced?

6. 2.1×10^3 L Chlorine in the upper atmosphere can destroy ozone molecules, O_3. The reaction can be represented by the following equation:

$$Cl_2(g) + 2O_3(g) \rightarrow 2ClO(g) + 2O_2(g)$$

How many liters of ozone can be destroyed at 220. K and 5.0 kPa if 200.0 g of chlorine gas react with it?

Name ______________________ Date ____________ Class ____________

CHAPTER 11 REVIEW

Molecular Composition of Gases

SECTION 11-4

SHORT ANSWER **Answer the following questions in the space provided.**

1. b, d, c, a List the following gases in order of increasing rate of effusion. (Assume all gases are at the same temperature and pressure.)

(a) He **(b)** Xe **(c)** HCl **(d)** Cl_2

2. c The two gases in the figure below are simultaneously injected into opposite ends of the tube. They should just begin to mix closest to which labeled point?

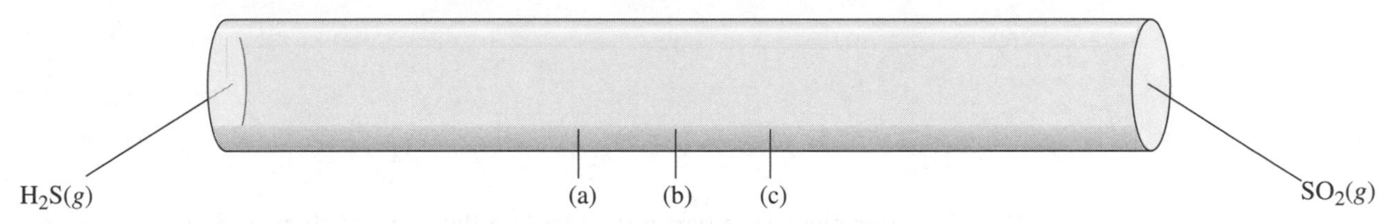

3. a In Graham's equation, the square roots of the molar masses can be substituted with the square roots of the ____.

(a) gas densities
(b) molar volumes
(c) compressibilities
(d) gas constants

4. f The ratio of the molar masses of two gases at the same temperature and pressure can be expressed by *r*. What will be the value of the ratio of their average molecular speeds?

(a) r
(b) r^2
(c) $\sqrt{r}$
(d) $\frac{1}{r}$
(e) $\left(\frac{1}{r}\right)^2$
(f) $\sqrt{\frac{1}{r}}$

5. Explain your reasoning for the order of gases you chose in item 1 above. Refer to the kinetic-molecular theory to support your explanation and cite Graham's law.

All gases at the same temperature have the same average kinetic energy. Therefore, heavier molecules have slower average speeds. Graham's law shows molecular speeds vary inversely with the square root of the molar masses. Thus, the gases are ranked from heaviest to lightest in molar mass.

PROBLEMS Write the answer on the line to the left. Show all your work in the space provided.

6. ___32 g/mol___ A gas of unknown molar mass is observed to effuse through a small hole at one-fourth the effusion rate of hydrogen. Estimate the molar mass of this gas.

7. ___1:9___ **a.** The molar masses of He and of HCl are 4.00 g/mol and 36.46 g/mol, respectively. What is the ratio of the mass of He to the mass of HCl rounded to one decimal place?

___1:3___ **b.** Use your answer in part a to calculate the ratio of their average speeds.

___400 m/s___ **c.** If helium's average speed is 1200 m/s, what is the average speed of HCl?

8. ___202 g/mol___ An unknown gas effuses through an opening at a rate 3.16 times slower than neon gas. Estimate the molar mass of this unknown gas.

Name ______________________ Date ______________ Class ______________

CHAPTER 11 REVIEW

Molecular Composition of Gases

MIXED REVIEW

SHORT ANSWER Answer the following questions in the space provided.

1. __c__ The average speed of a gas molecule is most directly related to the ____.

(a) polarity of the molecule
(b) pressure of the gas
(c) temperature of the gas
(d) number of moles in the sample

2. __d__ Which of the following statements best represents Graham's law for two gases at the same temperature?

(a) The molar masses and the velocities are inversely proportional.
(b) The molar masses and the velocities are directly proportional.
(c) The molar masses and the square roots of the velocities are directly proportional.
(d) The molar masses and the squares of the velocities are inversely proportional.

3. Water is synthesized from its elements according to the following equation:

$$2H_2(g) + O_2(g) \rightarrow 2H_2O(l)$$

__volume of H_2__ **a.** Which is greater, the volume of H_2 consumed or the volume of O_2 consumed?

__mass of O_2__ **b.** Which is greater, the mass of H_2 consumed or the mass of O_2 consumed?

4. Consider a 2.24 L sample of an ideal gas at STP.

__0.100 mol__ **a.** How many moles are present in the sample, or is more data needed?

__6.0×10^{22} molecules__ **b.** How many molecules are present in the sample, or is more data needed?

__more data is needed__ **c.** What is the mass of the sample, or is more data needed?

PROBLEMS Write the answer on the line to the left. Show all your work in the space provided.

5. __85 L__ Calculate the volume of 3.5 mol of an ideal gas at 24°C and 760 torr.

6. 60.0 g/mol ______ What is the molar mass of a gas that has a density of 2.68 g/L at STP?

7. A sample of oxygen gas has a mass of 8.00 g. Its boiling point is −183°C.

90 K ______ **a.** What is the boiling point in kelvins?

O_2; 32.00 g/mol ______ **b.** What is the formula and molar mass of oxygen?

0.25 mol ______ **c.** How many moles of oxygen are present in the 8.0 g sample?

0.80 atm ______ **d.** What pressure does the sample exert if it is placed in a rigid 8.21 L bottle at 77°C?

8. Hydrogen gas can be collected in the laboratory by adding an acid to aluminum metal according to the following equation:

$$2Al(s) + 6HCl(aq) \rightarrow 2AlCl_3(aq) + 3H_2(g)$$

12.2 L ______ **a.** Determine the volume of H_2 that forms at 20°C and a barometric pressure of 750. torr when 9.00 g of aluminum react according to the above equation.

12.5 L ______ **b.** Typically, hydrogen gas is collected by water displacement. Recalculate the volume of the dry gas taking into account the partial pressure of water vapor at the reaction temperature.

Name ______________________ Date ______________ Class ______________

CHAPTER 12 REVIEW

Liquids and Solids

SECTION 12-1

SHORT ANSWER **Answer the following questions in the space provided.**

1. b < d < a < c List the following attractive forces in order of increasing strength:

(a) hydrogen bonding
(b) London dispersion forces
(c) ionic bonding
(d) dipole-dipole forces

2. b All of the following statements about liquids and gases are true *except* ____.

(a) Molecules in a liquid are much more closely packed than molecules in a gas.
(b) Molecules in a liquid can vibrate and rotate, but they cannot move about freely as molecules in a gas.
(c) Liquids are much more difficult to compress into a smaller volume than are gases.
(d) Liquids diffuse more slowly than gases.

3. a Liquids posses all the following properties *except* ____.

(a) relatively low density
(b) the ability to diffuse
(c) relative incompressibility
(d) the ability to change to a gas

4. a. Chemists distinguish between intermolecular and intramolecular forces. Explain the difference between these two types of forces.

Intermolecular forces are between molecules; intramolecular forces are within molecules.

Classify each of the following as intramolecular or intermolecular:

intermolecular **b.** hydrogen bonding in liquid water

intramolecular **c.** the O—H covalent bond in methanol, CH_3OH

intermolecular **d.** the bonds that cause gaseous Cl_2 to become a liquid when cooled

5. Explain at the molecular level the following properties of liquids:

a. A liquid takes the shape of its container but does not expand to fill its volume.

Liquid molecules are very mobile. This mobility allows a liquid to take the shape of its container. In liquids, molecules are in contact with adjacent molecules, allowing intermolecular forces to have a greater effect than they do in gases. The molecules in a liquid will therefore not spread out to fill a container's volume.

Name ______________________ Date ______________ Class ______________

SECTION 12-1 continued

b. Polar liquids are slower to evaporate than nonpolar liquids.

Polar molecules are attracted to adjacent molecules and are therefore less able to escape from the liquid's surface than are nonpolar molecules.

c. Most liquids are much denser than their corresponding gases.

Most of a gas is empty space, with molecules far apart. In a liquid, molecules are in close contact with each other and are more closely packed.

6. Explain briefly why liquids tend to form spherical droplets, decreasing surface area to the smallest size possible.

Surface tension draws molecules inward. A sphere offers the minimum surface area for a given volume of liquid.

7. Is freezing a chemical change or a physical change? Explain your answer.

Freezing is a physical change. Even though the substance solidifying is changing its state, it is still the same substance.

Name ______________________ Date ______________ Class ______________

CHAPTER 12 REVIEW

Liquids and Solids

SECTION 12-2

SHORT ANSWER **Answer the following questions in the space provided.**

1. Match the following descriptions on the right to the crytal type on the left.

b	ionic crystal	(a) has mobile electrons in the crystal
c	covalent molecular crystal	(b) is hard, brittle, and nonconducting
a	metallic crystal	(c) typically has the lowest melting point of the four crystal types
d	covalent network crystal	(d) has strong covalent bonds between neighboring atoms

2. For each of the four types of solids, give a specific example other than those listed in Table 12-1 on page 370 of the text.

some possible answers: ionic solid: MgO, CaO, KI, $CuSO_4$

covalent network solid: graphite, silicon carbide

covalent molecular solid: dry ice (CO_2), sulfur, iodine

metallic solid: any metal from the left side of the periodic table

3. A chunk of solid lead is dropped into a pool of molten lead. The chunk sinks to the bottom of the pool. What does this tell you about the density of the solid lead compared with the density of the molten lead?

Solid lead is denser than the liquid form.

4. Answer *solid* or *liquid* to the following questions:

solid **a.** Which is more incompressible?

liquid **b.** Which is quicker to diffuse into neighboring media?

solid **c.** Which has a definite volume and shape?

solid **d.** Which has molecules that are primarily rotating or vibrating in place?

5. Explain at the molecular level the following properties of solids:

a. Solid metals conduct electricity well, but network solids do not.

Metals have many electrons that are not bound to any one atom; therefore they are able to move under an electric current. In network solids, all atoms (and electrons) are strongly bound in place and are not free to move.

b. Almost all solids are denser than their liquid state.

In solids, molecules are more closely packed than in liquids, causing the attractive forces between them to be at a maximum.

c. Amorphous solids do not have a definite melting point.

In amorphous solids, particles are arranged randomly; no specific amount of kinetic energy is needed to overcome the attractive forces holding the particles together.

d. Ionic crystals are much more brittle than covalent molecular crystals.

Ionic crystals have strong binding forces between the positive and negative ions in the crystal structure. Covalent molecular solids have weaker bonds between units.

6. Experiments show that it takes 6.0 kJ of heat energy to melt 1 mol of ice at its melting point but only about 0.6 kJ to melt 1 mol of methane, CH_4, at its melting point. Explain in terms of intermolecular forces why it takes so much less energy to melt the methane.

The attractive forces between CH_4 molecules are weak (dispersion forces). Little energy is needed to separate the molecules. Melting ice involves the breaking of many hydrogen bonds between molecules, which requires more energy.

Name ______________________ Date ____________ Class ____________

CHAPTER 12 REVIEW

Liquids and Solids

SECTION 12-3

SHORT ANSWER **Answer the following questions in the space provided.**

1. Consider the following system at equilibrium:

$$\text{reactants} \rightleftarrows \text{products}$$

A change in conditions causes the reverse reaction to be favored.

a. What happens to the concentration of the reactants?

The reactant concentration increases.

b. What happens to the concentration of the products?

The product concentration decreases.

2. The molar heat of vaporization of methane, CH_4, is 8.19 kJ/mol; for water, it is 40.79 kJ/mol.

3.0 kJ **a.** If 2.0×10^{23} molecules of liquid CH_4 are made to boil, how much heat must be supplied? Show all your work.

CH_4 **b.** Based on the molar heat of vaporization data, which is more volatile, CH_4 or H_2O?

H_2O **c.** Which molecule is more polar, CH_4 or H_2O?

3. A general equilibrium equation for boiling is:

$$\text{liquid} + \text{heat energy} \rightleftarrows \text{vapor}$$

Is the forward or reverse reaction favored in each of the following cases?

forward reaction **a.** The temperature of the system is increased.

reverse reaction **b.** More molecules of the vapor are added to the system.

forward reaction **c.** The pressure on the system is increased.

4. Consider water boiling in an open pot on a stove.

a. Can this system reach equilibrium? Why or why not?

It cannot reach equilibrium because the system is not closed.

b. Explain the difference between an open system and a closed system.

In a closed system, matter cannot enter or leave, but energy can. Both matter and energy can escape or enter an open system.

5. Methanol has a normal boiling point of 65°C. It is a liquid at conditions of 1 atm and 25°C. A small beaker filled with methanol is placed under a bell jar, and the air is then pumped out. It is observed that under a vacuum the methanol boils readily at 25°C.

Use the kinetic-molecular theory and the concept of equilibrium vapor pressure to account for the lowered boiling point of methanol under a vacuum.

As pressure on the system decreases, molecules of liquid at the surface require less energy to escape into the vapor phase. As pressure decreases, the amount of heat energy necessary for boiling to occur also decreases. When the pressure in the bell jar equals the vapor pressure of methanol at 25°C, the methanol boils.

6. Refer to the phase diagram for water in Figure 12-14 on page 381 of the text to answer the following questions:

A **a.** Which point represents the conditions under which all three phases can coexist?

C **b.** Which point represents a temperature above which only the solid phase exists?

decreases **c.** Based on the diagram, as the pressure on the water system is increased, the melting point of ice ____ (increases, decreases, or stays the same).

Name ______________________ Date ______________ Class ______________

CHAPTER 12 REVIEW

Liquids and Solids

SECTION 12-4

SHORT ANSWER **Answer the following questions in the space provided.**

1. Refer to the graph below to answer the following questions:

about 75°C **a.** What is the normal boiling point of CCl_4?

about 85°C **b.** What would be the boiling point of water if the air pressure over the liquid were reduced to 60 kPa?

about 40 kPa **c.** What must the air pressure over CCl_4 be for it to boil at 50°C?

d. Although water has a lower molar mass than CCl_4, it has a lower vapor pressure when measured at the same temperature. Why do you think water vapor is less volatile than CCl_4?

Highly polar molecules tend to attract each other when they are in the liquid state due to intermolecular forces, making evaporation more difficult. The symmetrical tetrahedral structure of CCl_4 makes it nonpolar; water is polar. Thus, water is less volatile despite its smaller molar mass.

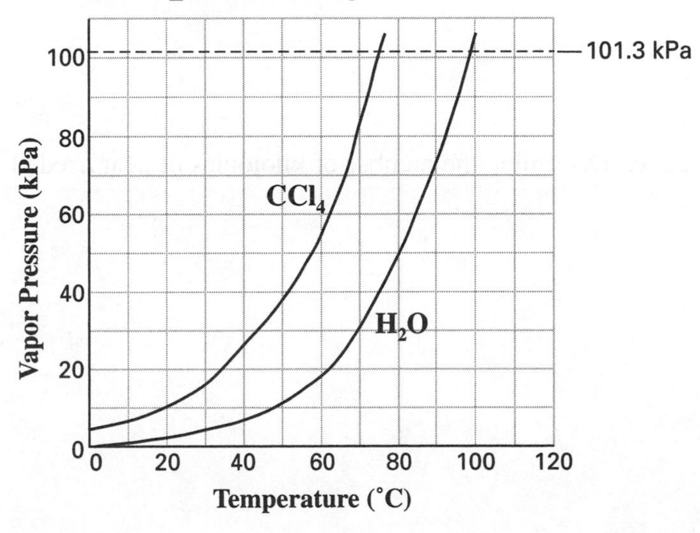

Name ______________________ Date ______________ Class ______________

SECTION 12-4 continued

2. Describe hydrogen bonding as it occurs in water in terms of the location of the bond, the particles involved, the strength of the bond, and the effects this type of bonding has on physical and chemical properties.

Hydrogen bonding in water occurs between a hydrogen atom of one water molecule and the unshared pair of electrons of an oxygen atom of an adjacent water molecule. It is a particularly strong type of dipole-dipole force. Hydrogen bonding causes the boiling point of water and its molar heat of vaporization to be relatively high. The phenomenon of surface tension is also a result of hydrogen bonding.

PROBLEM **Write the answer on the line to the left. Show all your work in the space provided.**

3. A typical ice cube has a volume of about 16 cm^3. Calculate the heat needed to melt the ice cube in a step by step manner. (Useful data: density of ice at 0.°C = 0.917 g/mL; heat of fusion of ice = 6.009 kJ/mol; heat of vaporization of H_2O = 40.79 kJ/mol; molar mass of H_2O = 18.02 g/mol)

15 g **a.** Determine the mass of the ice cube.

0.83 mol **b.** Determine the number of moles of H_2O present in the sample.

5.0 kJ **c.** Determine the number of kilojoules of heat needed to melt the ice cube.

Name ______________________ Date ______________ Class ______________

CHAPTER 12 REVIEW

Liquids and Solids

MIXED REVIEW

SHORT ANSWER **Answer the following questions in the space provided.**

1. Use this general equilibrium equation to answer the following questions:

reactants $\rightleftarrows$ products + heat energy

____decrease____ **a.** If the reaction shifts to the right, will the concentration of reactants increase, decrease, or stay the same?

____reverse reaction____ **b.** If extra product is introduced, which reaction will be favored?

____forward reaction____ **c.** If the temperature of the system decreases, which reaction will be favored?

2. Compare a polar water molecule with a less-polar molecule, such as formaldehyde, CH_2O. Both are liquids at room temperature and 1 atm pressure.

____water____ **a.** Which liquid should have the higher boiling point?

____formaldehyde____ **b.** Which liquid is more volatile?

____water____ **c.** Which liquid has a higher surface tension?

____formaldehyde____ **d.** Which liquid diffuses more rapidly?

____water____ **e.** In which liquid is NaCl, an ionic crystal, likely to be more soluble?

3. Use the data table below to answer the following:

Composition	Molar mass (g/mol)	Heat of vaporization (kJ/mol)	Normal boiling point (°C)	Critical temperature (°C)
He	4	0.08	−269	−268
Ne	20	1.8	−246	−229
Ar	40	6.5	−186	−122
Xe	131	12.6	−107	+17
H_2O	18	40.8	+100	+374
HF	20	25.2	+20	+188
CH_4	16	8.9	−161	−82
C_2H_6	30	15.7	−89	+32

____increase____ **a.** Among nonpolar liquids, as their molar mass increases, their normal boiling point tends to ____ (increase, decrease, or stay about the same).

MIXED REVIEW continued

decrease **b.** Among compounds of approximately the same molar mass, as their polarity increases, their heat of vaporization tends to ____ (increase, decrease, or stay about the same).

c. Which is the only noble gas listed that is stable as a liquid at 0°C? Explain your answer, using the concept of critical temperature.

Xe; A substance can exist only as a gas at temperatures above its critical temperature. Of the noble gases listed, only Xe has a critical temperature above 0°C.

PROBLEMS Write the answer on the line to the left. Show all your work in the space provided.

4. The heat of fusion of ice is 6.009 kJ/mol.

0.33 kJ **a.** How much heat is needed to melt 1.0 g of ice?

80. cal/g **b.** An energy unit often encountered is the calorie (4.18 J = 1 calorie). Determine the heat of fusion of ice in calories/gram.

5. 181 kJ Freon-11, CCl_3F, has been commonly used in air conditioners. Its heat of vaporization is 24.8 kJ/mol at its normal boiling point of 24°C. How much heat is removed from a room by an air conditioner that evaporates 1.00 kg of freon-11?

Name ______________________ Date ______________ Class ______________

CHAPTER 13 REVIEW

Solutions

SECTION 13-1

SHORT ANSWER **Answer the following questions in the space provided.**

1. Match the type of mixture on the left to its representative particle diameter on the right.

c solutions (a) larger than 1000 nm

a suspensions (b) 1 nm to 1000 nm

b colloids (c) smaller than 1 nm

2. Identify the solvent in each of the following examples:

alcohol **a.** tincture of iodine (iodine dissolved in ethyl alcohol)

water **b.** sea water

the gel **c.** water-absorbing super gels

3. A mixture has the following properties:

- No solid settles out during a 48 hour period.
- The path of a flashlight beam is easily seen through the mixture.
- It appears to be homogeneous under a hand lens but not under a microscope.

Is the mixture a suspension, colloid, or true solution? Explain your answer.

The mixture is a colloid. The properties are consistent with those reported in Table 13-3 on page 398 of the text. The particle size is small, but not too small, and the mixture exhibits theTyndall effect.

4. Define each of the following terms:

a. alloy

a homogeneous mixture of two or more solid metals

b. electrolyte

a substance that dissolves in water to form a solution that conducts an electric current

c. aerosol

a colloidal dispersion of a solid in a gas

d. aqueous solution

a mixture with a soluble solute and water as the solvent

5. For each of the following types of solutions, give an example other than those listed in Table 13-1 on page 396 of the text:

a. a gas in a liquid

oxygen gas dissolved in water (needed by fish)

b. a liquid in a liquid

antifreeze, which is ethylene glycol dissolved in water

c. a solid in a liquid

salt dissolved in water or iodine in alcohol

6. Of the following solution models shown at the particle level, indicate which will conduct electricity. Give reasons for your answers.

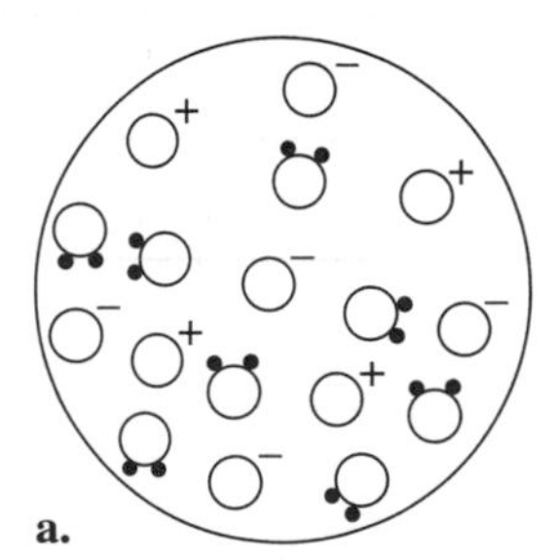

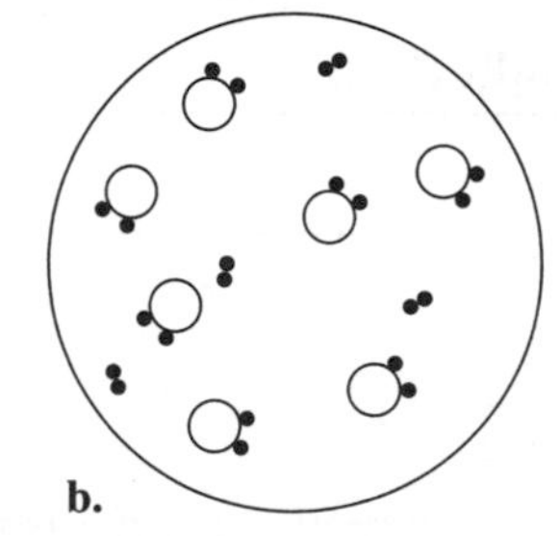

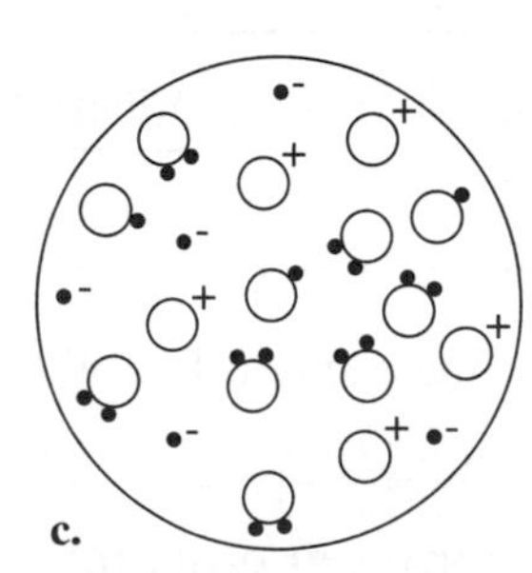

a. **will conduct electricity because it dissociates into ions**

b. **will not conduct electricity because it does not dissociate into ions**

c. **will conduct electricity because it dissociates into ions**

Name ______________________ Date ______________ Class ______________

CHAPTER 13 REVIEW

Solutions

SECTION 13-2

SHORT ANSWER **Answer the following questions in the space provided.**

1. Following are statements about the dissolving process. Explain each one at the molecular level.

a. Increasing the pressure of a solute gas above a liquid solution increases the solubility of the gas in the liquid.

Increasing the pressure of the solute gas above the solution puts stress on the equilibrium of the system. Gas molecules collide with the liquid surface more often, causing an increase in the rate of gas molecules entering into solution.

b. Increasing the temperature of water speeds up the rate at which many solids dissolve in this solvent.

As the temperature of the water increases, water molecules move faster, increasing their average kinetic energy. At higher temperatures, collisions between the water molecules and the solute are more frequent and are of higher energy than at lower temperatures. This helps to separate solute particles from one another and to disperse them among the water molecules.

c. Increasing the surface area of a solid solute speeds up the rate at which it dissolves in a liquid solvent.

Increasing the surface area of a solid exposes more of the solute to the solvent, allowing the solvent to come into contact with more of the solute in a shorter length of time.

2. The solubility of $KClO_3$ at 25°C is 10. g of solute per 100. g of H_2O.

a. If 15 g of $KClO_3$ are added to 100 g of water at 25°C with stirring, how much of the $KClO_3$ will dissolve? Is the solution saturated, unsaturated, or supersaturated?

10 g of $KClO_3$ will dissolve, but 5 g will not, despite thorough stirring. The solution is saturated.

b. If 15 g of $KClO_3$ are added to 200 g of water at 25°C with stirring, how much of the $KClO_3$ will dissolve? Is the solution saturated, unsaturated, or supersaturated?

All 15 g of $KClO_3$ will dissolve; the solution is unsaturated.

PROBLEMS **Write the answer on the line to the left. Show all your work in the space provided.**

3. Use the data in Table 13-4 on page 404 of the text to answer the following questions:

251 g **a.** How many grams of LiCl are needed to make a saturated solution with 300. g of water at 20°C?

50. g **b.** What is the minimum amount of water needed to dissolve 51 g of $NaNO_3$ at 40°C?

KI **c.** Which salt forms a saturated solution when 36 g of it are dissolved in 25 g of water at 20°C?

4. KOH is an ionic solid readily soluble in water.

−1.027 kJ/g **a.** What is its heat of solution in kJ/g? Refer to the data in Table 13-5 on page 410 of the text.

b. Will the temperature of the system increase or decrease as the dissolution of KOH proceeds? Why?

The temperature of the system will increase, because the reaction is exothermic, giving off heat energy to warm up the system.

Name ______________________ Date ______________ Class ______________

CHAPTER 13 REVIEW

Solutions

SECTION 13-3

SHORT ANSWER **Answer the following questions in the space provided.**

1. Describe the errors made by the following students in making molar solutions.

a. James needs a 0.600 M solution of KCl. He measures out 0.600 g of KCl and adds 1 L of water to the solid.

James made many errors. 0.600 mol does not have a mass of 0.600 g. Also, adding 1.0 L of water to the solid does not produce 1.0 L of solution. He did not make a 0.600 M solution.

b. Mary needs a 0.02 M solution of $NaNO_3$. She calculates that she needs 2.00 g of $NaNO_3$ for 0.02 mol. She puts this solid into a 1.00 L volumetric flask and fills the flask to the 1.00 L mark.

Mary did not produce the required solution either. 0.02 mol of $NaNO_3$ has a mass of 1.70 g, not 2.00 g. Also, she should have made sure the solute was completely dissolved before continuing to fill the volumetric flask to the mark.

PROBLEMS **Write the answer on the line to the left. Show all your work in the space provided.**

2. **0.33 M** What is the molarity of a solution made by dissolving 2.0 mol of solute in 6.0 L of solvent?

3. **1.0 *m*** CH_3OH is soluble in water. What is the molality of a solution made by dissolving 8.0 g of CH_3OH in 250. g of water?

4. Marble chips effervesce when treated with acid. This reaction is represented by the following equation:

$$CaCO_3(s) + 2HCl(aq) \rightarrow CaCl_2(aq) + CO_2(g) + H_2O(l)$$

To produce a reaction, 25.0 mL of 4.0 M HCl is added to excess $CaCO_3$.

0.10 mol **a.** How many moles of HCl are consumed in this reaction?

1.1 L **b.** How many liters of CO_2 are produced at STP?

5.0 g **c.** How many grams of $CaCO_3$ are consumed?

5. Tincture of iodine is $I_2(s)$ dissolved in ethanol, C_2H_5OH. A 1% solution of tincture of iodine is 10.0 g of solute for 1000. g of solution.

990. g **a.** How many grams of solvent are present in 1000. g of this solution?

0.0394 mol **b.** 10.0 g of I_2 represent how many moles of solute?

0.0398 *m* **c.** What is the molality of this 1% solution?

d. To determine a solution's molarity, the density of that solution can be used. Explain how you would use the density of the tincture of iodine solution to calculate its molarity.

The density of a solution can be expressed in g/mL or in kg/L. Divide 1.00 kg by the solution's density to find the volume of solution in liters. Then divide 0.0394 mol by this volume to arrive at the molarity.

Name ____________________ Date ______________ Class ______________

CHAPTER 13 REVIEW

Solutions

MIXED REVIEW

SHORT ANSWER **Answer the following questions in the space provided.**

1. Solid $CaCl_2$ does not conduct electricity, but it is considered to be an electrolyte. Explain.

In the crystal form, ions are locked in position. Dissolving the crystal in water, however, releases the ions to move freely, thereby able to conduct electricity.

2. Explain the following statements at the molecular level:

a. Generally a polar liquid and a nonpolar liquid are immiscible.

Polar molecules tend to attract one another, forcing the nonpolar molecules to remain in a separate layer.

b. Carbonated soft drinks taste flat when they are warmed.

The solubility of gases decreases as the temperature of the solution increases. At higher temperatures, more CO_2 molecules escape through the liquid's surface, leaving fewer molecules in solution to effervesce.

3. An unknown compound is observed to mix with toluene, $C_6H_5CH_3$, but not with water.

a. Is the unknown compound ionic, polar covalent, or nonpolar covalent?

nonpolar covalent, because it mixes well with nonpolar toluene

b. Suppose the unknown compound is also a liquid. Will it be able to dissolve table salt? Explain your answer.

No; being nonpolar, the solvent molecules are unable to remove ions from crystal surfaces.

PROBLEMS **Write the answer on the line to the left. Show all your work in the space provided.**

4. Consider 500. mL of a 0.30 M $CuSO_4$ solution.

0.15 mol **a.** How many moles of solute are present in this solution?

24 g **b.** How many grams of solute were used to prepare this solution?

5. **a.** If a solution is electrically neutral, can all of its ions have the same charge? Explain your answer.

No; to be neutral the total positive charge must equal the total negative charge.

6.0×10^{13} ions **b.** The concentration of the OH^- ions in pure water is known to be 1.0×10^{-7} M. How many OH^- ions are present in each milliliter of pure water?

6. 90. g of $CaBr_2$ are dissolved in 900. g of water.

900. mL **a.** What volume does the 900. g of water occupy if its density is 1.00 g/mL?

0.50 *m* **b.** What is the molality of this solution?

Name ____________________ Date ____________ Class ____________

CHAPTER 14 REVIEW

Ions in Aqueous Solutions and Colligative Properties

SECTION 14-1

SHORT ANSWER **Answer the following questions in the space provided.**

1. Use the guidelines in Table 14-1 on page 427 of the text to predict the solubility of the following compounds in water:

soluble **a.** magnesium nitrate

insoluble **b.** barium sulfate

insoluble **c.** calcium carbonate

soluble **d.** ammonium phosphate

2. 1.0 mol of magnesium acetate is dissolved in water.

$Mg(CH_3COO)_2$ **a.** Write the formula for magnesium acetate.

3.0 mol **b.** How many moles of ions are released into solution?

0.60 mol **c.** How many moles of ions are released into a solution made from 0.20 mol magnesium acetate dissolved in water?

3. In the following two precipitation reactions, write the formula for the precipitate formed:

$Mg_3(PO_4)_2$ **a.** combining solutions of magnesium chloride and potassium phosphate

Ag_2S **b.** combining solutions of sodium sulfide and silver nitrate

4. Write ionic equations showing the dissolution of the following compounds:

a. $Na_3PO_4(s)$

$Na_3PO_4(s) \rightarrow 3Na^+(aq) + PO_4^{3-}(aq)$

b. iron(III) sulfate(*s*)

$Fe_2(SO_4)_3(s) \rightarrow 2Fe^{3+}(aq) + 3SO_4^{2-}(aq)$

5. a. Write the net ionic equation for the reaction that occurs when solutions of lead(II) nitrate and ammonium sulfate are combined.

$Pb^{2+}(aq) + SO_4^{2-}(aq) \rightarrow PbSO_4(s)$

b. What are the spectator ions in this system?

NO_3^- and NH_4^+ are spectator ions.

Name ______________________ Date ______________ Class ______________

SECTION 14-1 continued

6. The following solutions are combined in a beaker: NaCl, Na_3PO_4, and $Ba(NO_3)_2$

a. Write the name of each of the compounds used.

sodium chloride, sodium phosphate, and barium nitrate

b. Will a precipitate form when the above solutions are combined? If so, write the name and formula of the precipitate.

Yes; barium phosphate, $Ba_3(PO_4)_2$, forms as a precipitate.

c. List all spectator ions present in this system.

Na^+, Cl^-, and NO_3^- are spectator ions in this system.

7. It is possible to have spectator ions present in many chemical systems, not just in precipitation reactions. For example:

$$Al(s) + HCl(aq) \rightarrow AlCl_3(aq) + H_2(g) \text{ (unbalanced)}$$

True **a.** In an aqueous solution of HCl, virtually every HCl molecule is ionized. True or False?

$Cl^-(aq)$ **b.** There is only one spectator ion in this system. Is it $Al^{3+}(aq)$, $H^+(aq)$, or $Cl^-(aq)$?

c. Balance the above equation.

$2Al(s) + 6HCl(aq) \rightarrow 2AlCl_3(aq) + 3H_2(g)$

11 L **d.** If 9.0 g of Al metal react with excess HCl according to the balanced equation in part c, what volume of hydrogen gas at STP will be produced? Show all your work.

Name ________________ Date ________________ Class ________________

CHAPTER 14 REVIEW

Ions in Aqueous Solutions and Colligative Properties

SECTION 14-2

PROBLEMS Write the answer on the line to the left. Show all your work in the space provided.

1. 100.102°C **a.** Predict the boiling point of a 0.200 *m* solution of glucose in water.

100.204°C **b.** Predict the boiling point of a 0.200 *m* solution of potassium iodide in water.

2. A chief ingredient of antifreeze is liquid ethylene glycol, $C_2H_4(OH)_2$. Assume $C_2H_4(OH)_2$ is added to a car radiator that holds 5.0 kg of water.

48 mol **a.** How many moles of ethylene glycol should be added to the water in the radiator to lower the freezing point of that water from 0°C to −18°C?

3.0×10^3 g **b.** How many grams of ethylene glycol does the quantity in part a represent?

2.7 L **c.** Ethylene glycol has a density of 1.1 kg/L. How many liters of $C_2H_4(OH)_2$ should be added to the water in the radiator to protect the system to −18°C?

SECTION 14-2 continued

d. In World War II, soldiers in the Sahara desert needed a supply of antifreeze to protect the radiators of their vehicles. Since the temperature in the Sahara almost never drops to 0°C, why was the antifreeze necessary?

Antifreeze also raises the boiling point of water. It was needed to help prevent the water in the radiators of the vehicles from boiling over.

3. An important use of colligative properties is to determine the molar mass of unknown substances. The following situation is an example: 12.0 g of unknown compound X, a nonpolar nonelectrolyte, is dissolved in 100.0 g of melted camphor. The resulting solution freezes at 99.4°C. Consult Table 14-2 on page 438 of the text for any other data needed to answer the following questions:

79.4°C **a.** By how many Celsius degrees did the freezing point of camphor change from its normal freezing point?

2.00 *m* **b.** What is the molality of the solution of camphor and compound X based on freezing point data?

120. g **c.** If there are 12.0 g of compound X per 100.0 g of camphor, how many grams of compound X are there per kilogram of camphor?

60.0 g/mol **d.** What is the molar mass of compound X?

Name ______________________ Date ______________ Class ______________

CHAPTER 14 REVIEW

Ions in Aqueous Solutions and Colligative Properties

MIXED REVIEW

SHORT ANSWER **Answer the following questions in the space provided.**

1. Match the four compounds on the right to their descriptions on the left.

b	an ionic compound that is quite soluble in water	**(a)** HCl
c	an ionic compound that is not very soluble in water	**(b)** $BaCl_2$
a	a molecular compound that ionizes in water	**(c)** AgCl
d	a molecular compound that does not ionize in water	**(d)** CCl_4

2. Consider nonelectrolytes dissolved in various liquid solvents to complete the following statements:

solute **a.** The change in the boiling point does not vary with the identity of the ____ (solute, solvent), assuming all other factors remain constant.

solvent **b.** The change in the boiling point does vary with the identity of the ____ (solute, solvent), assuming all other factors remain constant.

increases **c.** The change in the boiling point becomes greater as the concentration of the solute in solution ____ (increases, decreases).

3. a. Name two compounds in solution that could be combined to cause calcium carbonate to precipitate.

Answers will vary; any soluble calcium salt mixed with any soluble carbonate will form the precipitate. One example is calcium nitrate with sodium carbonate.

b. Identify any spectator ions in the system you described in part a.

In the example given, sodium and nitrate ions are spectator ions.

c. Write the net ionic equation for the formation of calcium carbonate.

$Ca^{2+}(aq) + CO_3^{2-}(aq) \rightarrow CaCO_3(s)$

4. Explain why applying rock salt (impure NaCl) to an icy sidewalk hastens the melting process.

The vapor pressure of the NaCl solution forming is lower than the vapor pressure of pure water at 0°C. The lower vapor pressure of the NaCl solution results in a lower freezing point.

MIXED REVIEW continued

PROBLEMS Write the answer on the line to the left. Show all your work in the space provided.

5. ___13.4 *m*___ Some insects survive cold winters by generating an antifreeze inside their cells. The antifreeze produced is glycerol, $C_3H_5(OH)_3$, a nonelectrolyte that is quite soluble in water. What must the molality of a glycerol solution be to lower the freezing point of water to $-25.0°C$?

6. ___2.14 g___ How many grams of methanol, CH_3OH, should be added to 200. g of acetic acid to lower its freezing point by 1.30°C? Refer to Table 14-2 on page 438 of the text for any necessary data.

7. ___0.67 *m*___ The boiling point of a solution of glucose, $C_6H_{12}O_6$, and water was recorded to be 100.34°C. Calculate the molality of this solution.

8. HF(*aq*) is a weak acid. A 0.05 mol sample of HF is added to 1.0 kg of water.

a. Write the equation for the ionization of HF to form hydronium ions.

$HF(aq) + H_2O(l) \rightarrow H_3O^+(aq) + F^-(aq)$

___0.10 mol___ **b.** If HF were 100% ionized, how many moles of its ions would be released?

Name ______ Date ______ Class ______

CHAPTER 15 REVIEW

Acids and Bases

SECTION 15-1

SHORT ANSWER Answer the following questions in the space provided.

1. Name the following compounds as acids:

sulfuric acid **a.** H_2SO_4

sulfurous acid **b.** H_2SO_3

hydrosulfuric acid **c.** H_2S

perchloric acid **d.** $HClO_4$

hydrocyanic acid **e.** hydrogen cyanide

2. H_2S Which (if any) of the acids mentioned in item 1 are binary acids?

3. Write formulas for the following acids:

HNO_2 **a.** nitrous acid

HBr **b.** hydrobromic acid

H_3PO_4 **c.** phosphoric acid

CH_3COOH **d.** acetic acid

HClO **e.** hypochlorous acid

4. Calcium selenate has the formula $CaSeO_4$.

H_2SeO_4 **a.** What is the formula for selenic acid?

H_2SeO_3 **b.** What is the formula for selenous acid?

5. Use an activity series to identify two metals that will not generate hydrogen gas when treated with an acid.

Choose from Cu, Ag, Au, Pt, Pd, or Hg.

6. Write balanced molecular equations for the following reactions of acids and bases:

a. aluminum metal with dilute nitric acid

$2Al(s) + 6HNO_3(aq) \rightarrow 2Al(NO_3)_3(aq) + 3H_2(g)$

b. calcium hydroxide solution with acetic acid

$Ca(OH)_2(aq) + 2CH_3COOH(aq) \rightarrow Ca(CH_3COO)_2(aq) + 2H_2O(l)$

SECTION 15-1 continued

7. Write net ionic equations that represent the following reactions:

a. the ionization of $HClO_3$ in water

$HClO_3(aq) + H_2O(l) \rightleftarrows H_3O^+(aq) + ClO_3^-(aq)$

b. NH_3 functioning as an Arrhenius base

$NH_3(aq) + H_2O(l) \rightleftarrows NH_4^+(aq) + OH^-(aq)$

8. a. Explain how strong acid solutions conduct an electric current.

Strong acids release many ions into solution. These ions are free to move, making it possible for an electric current to pass through the solution.

b. Will a strong acid or a weak acid conduct electricity better, assuming all other factors remain constant? Explain your answer.

Strong acids conduct electricity better because they have many more ions present per liter of solution than do weak acids of the same concentration.

9. Most acids react with solid carbonates. For example:

$$CaCO_3(s) + HCl(aq) \rightarrow CaCl_2(aq) + H_2O(l) + CO_2(g) \text{ (unbalanced)}$$

a. Balance the above equation.

$CaCO_3(s) + 2HCl(aq) \rightarrow CaCl_2(aq) + H_2O(l) + CO_2(g)$

b. Write the net ionic equation for the above reaction.

$CO_3^{2-}(aq) + 2H^+(aq) \rightarrow H_2O(l) + CO_2(g)$

Ca^{2+} and Cl^- **c.** Identify all spectator ions in this system.

1.1 L **d.** How many liters of CO_2 form at STP if 5.0 g of $CaCO_3$ are treated with excess hydrochloric acid? Show all your work.

Name ______________________ Date ______________ Class ____________

CHAPTER 15 REVIEW

Acids and Bases

SECTION 15-2

SHORT ANSWER Answer the following questions in the space provided.

1. **a.** Write the two equations that show the two-stage ionization of sulfurous acid in water.

stage 1: $H_2SO_3(aq) + H_2O(l) \rightleftarrows H_3O^+(aq) + HSO_3^-(aq)$

stage 2: $HSO_3^-(aq) + H_2O(l) \rightleftarrows H_3O^+(aq) + SO_3^{2-}(aq)$

b. Which stage of ionization is likely favored?

stage 1: for most polyprotic acids, the concentration of ions formed in the first ionization is the greatest.

2. **a.** Define a Lewis base. Can OH^- function as a Lewis base? Explain your answer.

A Lewis base is a species that donates an electron pair to form a covalent bond. Yes, OH^- is a Lewis base. There are two nonbonding electron pairs available for it to donate. For example: $OH^-(aq) + Al(OH)_3(aq) \rightarrow Al(OH)_4^-(aq)$

b. Define a Lewis acid. Can H^+ function as a Lewis acid? Explain your answer.

A Lewis acid is a species that accepts an electron pair into a vacant orbital to form a covalent bond. H^+ is a Lewis acid. Its 1*s* orbital is vacant, and an electron pair from a base can fill it. For example: $H^+(aq) + OH^-(aq) \rightarrow H_2O(l)$

3. **a.** Identify the Brønsted-Lowry acid and the Brønsted-Lowry base on the reactant side of each of the following reactions, which occur in aqueous solution. Explain your answers.

a. $H_2O(l) + HNO_3(aq) \rightarrow H_3O^+(aq) + NO_3^-(aq)$

HNO_3 is the Brønsted-Lowry acid because it donates a proton to the H_2O. The H_2O is the Brønsted-Lowry base because it is the proton acceptor.

b. $HF(aq) + HS^-(aq) \rightarrow H_2S(aq) + F^-(aq)$

HF is the Brønsted-Lowry acid because it donates a proton to the HS^-. The HS^- is the Brønsted-Lowry base because it is the proton acceptor.

SECTION 15-2 continued

4. a. Write the equation for the first ionization of H_2CO_3 in aqueous solution. Assume that water serves as the reactant that attaches to the hydrogen ion released from the H_2CO_3. Which of the reactants is the Brønsted-Lowry acid and which is the Brønsted-Lowry base? Explain your answer.

$H_2CO_3(aq) + H_2O(l) \rightarrow HCO_3^-(aq) + H_3O^+(aq)$. The H_2CO_3 is the Brønsted-Lowry acid because it donates a proton to the H_2O. The H_2O is the Brønsted-Lowry base because it accepts the proton.

b. Write the equation for the second ionization, that of the ion that was formed by the H_2CO_3 in the equation you just wrote. Again, assume that water serves as the reactant that attaches to the hydrogen ion released. Which of the reactants is the Brønsted-Lowry acid and which is the Brønsted-Lowry base? Explain your answer.

$HCO_3^-(aq) + H_2O(l) \rightarrow CO_3^{2-}(aq) + H_3O^+(aq)$. The HCO_3^- is the Brønsted-Lowry acid because it donates a proton to the H_2O. The H_2O is the Brønsted-Lowry base because it accepts the proton.

c. What is the name for a substance, such as H_2CO_3, that can donate two protons?

a diprotic acid

5. a. How many electron pairs surround an atom of boron (B, element 5) bonded in the compound BCl_3?

three

b. How many electron pairs surround an atom of nitrogen (N, element 7) in the compound NF_3?

four

c. Write an equation for the reaction between the two compounds above. Assume that they react in a 1:1 ratio to form one molecule as product.

$BCl_3 + NF_3 \rightarrow BCl_3NF_3$

d. Assuming that the B and the N are covalently bonded to each other in the product, which of the reactants is the Lewis acid? Is this reactant also a Brønsted-Lowry acid? Explain your answers.

BCl_3 is the Lewis acid because it accepts an electron pair in forming a covalent bond. It is not a Brønsted-Lowry acid, because it is not donating a proton.

e. Which of the reactants is the Lewis base? Explain your answer.

NF_3 is the Lewis base because it donates an electron pair in forming a covalent bond.

Name ______________________ Date ______________ Class ______________

CHAPTER 15 REVIEW

Acids and Bases

SECTION 15-3

SHORT ANSWER **Answer the following questions in the space provided.**

1. Answer the following questions according to the Brønsted-Lowry definitions of acids and bases:

HSO_3^- **a.** What is the conjugate base of H_2SO_3?

NH_3 **b.** What is the conjugate base of NH_4^+?

OH^- **c.** What is the conjugate base of H_2O?

H_3O^+ **d.** What is the conjugate acid of H_2O?

$H_2AsO_4^-$ **e.** What is the conjugate acid of $HAsO_4^{2-}$?

2. Consider the following reaction:

$$NH_4^+(aq) + CO_3^{2-}(aq) \rightleftarrows NH_3(aq) + HCO_3^-(aq)$$

a. If NH_4^+ is labeled as *acid 1*, label the other three terms as *acid 2*, *base 1*, and *base 2* to indicate the conjugate acid-base pairs.

base 2 CO_3^{2-}

acid 2 HCO_3^-

base 1 NH_3

True **b.** A proton has been transferred from acid 1 to base 2 in the above reaction. True or False?

3. Given the following neutralization reaction: $HCO_3^-(aq) + OH^-(aq) \rightleftarrows CO_3^{2-}(aq) + H_2O(l)$

a. Label the conjugate acid-base pairs in this system.

$HCO_3^-(aq) + OH^-(aq) \rightleftarrows CO_3^{2-}(aq) + H_2O(l)$

acid 1 base 2 base 1 acid 2

b. Is the forward or reverse reaction favored? Explain your answer.

the forward reaction; The weaker acid and weaker base are produced in the forward reaction. HCO_3^- competes more strongly with H_2O to donate a proton and OH^- competes more strongly with CO_3^{2-} to acquire a proton, causing the forward reaction to be favored.

SECTION 15-3 continued

4. Table 15-6 on page 471 of the text lists several amphoteric species but only one other than water is neutral.

NH_3 **a.** Identify that neutral compound.

b. Write two equations that demonstrate this compound's amphoteric properties.

(Answers will vary, but one equation should form NH_4^+, the other NH_2^-.)

$NH_3(aq) + HCl(aq) \rightarrow NH_4^+(aq) + Cl^-(aq)$

$NH_3(aq) + H^-(aq) \rightarrow NH_2^-(aq) + H_2(g)$

5. Write the formula for the salt formed in each of the following neutralization reactions:

K_3PO_4 **a.** potassium hydroxide combined with phosphoric acid

$Ca(NO_2)_2$ **b.** calcium hydroxide combined with nitrous acid

$BaBr_2$ **c.** hydrobromic acid combined with barium hydroxide

Li_2SO_4 **d.** lithium hydroxide combined with sulfuric acid

6. Given the following unbalanced equation for a neutralization reaction:

$$H_2SO_4(aq) + NaOH(aq) \rightarrow Na_2SO_4(aq) + H_2O(l)$$

a. Balance the equation.

$H_2SO_4(aq) + 2NaOH(aq) \rightarrow Na_2SO_4(aq) + 2H_2O(l)$

Na^+ and SO_4^{2-} **b.** In this system there are two spectator ions. Identify them.

1:2 **c.** In order to completely consume all reactants, what should be the mole ratio of acid to base?

7. The gases that produce acid rain are often referred to as NO_x and SO_x.

a. List three specific examples of these gases.

Some examples are NO, NO_2, N_2O_3, N_2O_5, SO_2, and SO_3.

b. Coal- and oil-burning power plants oxidize any sulfur in their fuel as it burns in air to form SO_2 gas. The SO_2 is further oxidized by O_2 in our atmosphere to form SO_3 gas. The SO_3 gas can combine with water to form sulfuric acid. Write balanced molecular equations to illustrate these three reactions.

$S(s) + O_2(g) \rightarrow SO_2(g)$

$2SO_2(g) + O_2(g) \rightarrow 2SO_3(g)$

$SO_3(g) + H_2O(l) \rightarrow H_2SO_4(aq)$

c. Industrial plants making fertilizers and detergents release NO_x gases into the air. Write a balanced equation for converting $N_2O_5(g)$ into nitric acid by reacting it with water.

$N_2O_5(g) + H_2O(l) \rightarrow 2HNO_3(aq)$

Name ______________________ Date ______________ Class ______________

CHAPTER 15 REVIEW

Acids and Bases

MIXED REVIEW

SHORT ANSWER **Answer the following questions in the space provided.**

1. **HClO** a. Write the formula for hypochlorous acid.

   hydrofluoric acid b. Write the name for HF(*aq*).

   $H_2C_2O_4$ c. If $Pb(C_2O_4)_2$ is lead(IV) oxalate, what is the formula for oxalic acid?

   acetic acid d. Name the acid that is present in vinegar.

2. Answer the following questions according to Brønsted-Lowry acid-base theory. Consult Table 15-5 on page 468 of the text as needed.

   HS^- a. What is the conjugate base of H_2S?

   HPO_4^{3-} b. What is the conjugate base of HPO_4^{2-}?

   NH_4^+ c. What is the conjugate acid of NH_3?

3. Consider the following equation:

$$OH^-(aq) + HCO_3^-(aq) \rightarrow H_2O(l) + CO_3^{2-}(aq)$$

   If OH^- is base 1, label the other three terms.

   H_2O a. acid 1

   HCO_3^- b. acid 2

   CO_3^{2-} c. base 2

4. Write the formula for the salt that is produced in the following neutralization reactions:

   K_2SO_3 a. sulfurous acid combined with potassium hydroxide

   $Ca_3(PO_4)_2$ b. calcium hydroxide combined with phosphoric acid

5. Carbonic acid releases H_3O^+ ions into water in two stages.

   a. Write equations representing each stage.

   stage 1: $H_2CO_3(aq) + H_2O(l) \rightleftarrows H_3O^+(aq) + HCO_3^-(aq)$

   stage 2: $HCO_3^-(aq) + H_2O(l) \rightleftarrows H_3O^+(aq) + CO_3^{2-}(aq)$

   stage 1 b. Which stage releases the most ions into solution?

MIXED REVIEW continued

6. Glacial acetic acid is a highly viscous liquid that is close to 100% CH_3COOH. When it mixes with water, it forms dilute acetic acid.

a. When making a dilute acid solution, should you add acid to water or water to acid? Explain your answer.

Add acid to water to achieve a thorough mixing of a denser acid with a slow release of heat and to avoid splashing concentrated acid.

b. Glacial acetic acid does not conduct electricity, but dilute acetic acid does. Explain your answer.

Glacial acetic acid exists as neutral molecules. In the presence of water, some of those molecules ionize into H^+ and CH_3COO^-, which can conduct electricity.

c. Dilute acetic acid does not conduct electricity as well as dilute nitric acid at the same concentration. Is acetic acid a strong or weak acid?

weak

d. Although there are four H atoms per molecule, acetic acid is monoprotic. Show the structural formula for CH_3COOH, and indicate the H atom that ionizes.

```
     H     O—H
     |    /
  H—C—C
     |    \\
     H     O
```

The H bonded to the O ionizes.

e. **30. g** How many grams of glacial acetic acid should be used to make 250 mL of 2.00 M acetic acid? Show all your work.

7. The overall effect of acid rain on lakes and ponds is partially determined by the geology of the lake bed. In some cases, the rock is limestone, which is rich in calcium carbonate. Calcium carbonate reacts with the acid in lake water according to the following (incomplete) ionic equation:

$$CaCO_3(s) + 2H_3O^+(aq) \rightarrow$$

a. Complete the ionic equation shown above.

$CaCO_3(s) + 2H_3O^+(aq) \rightarrow Ca^{2+}(aq) + CO_2(g) + 3H_2O(l)$

b. As this reaction occurs, does the concentration of H_3O^+ in the lake water increase or decrease? What effect does this have on the acidity of the lake water?

decreases, making the lake water less acidic

Name ________________ Date ________________ Class ________________

CHAPTER 16 REVIEW

Acid-Base Titration and pH

SECTION 16-1

SHORT ANSWER Answer the following questions in the space provided.

1. Calculate the following values: (A calculator should not be necessary.)

1×10^{-8} M **a.** If the $[H_3O^+] = 1 \times 10^{-6}$ M for a solution, calculate the $[OH^-]$.

1×10^{-5} M **b.** If the $[H_3O^+] = 1 \times 10^{-9}$ M for a solution, calculate the $[OH^-]$.

1×10^{-2} M **c.** If the $[OH^-] = 1 \times 10^{-12}$ M for a solution, calculate the $[H_3O^+]$.

2×10^{-2} M **d.** If the $[OH^-]$ in part c is reduced by half, to 0.5×10^{-12} M, calculate the $[H_3O^+]$.

inversely **e.** The $[H_3O^+]$ and $[OH^-]$ are ____ (directly, inversely, or not) proportional in any system involving water.

2. Calculate the following values: (A calculator should not be necessary.)

12.0 **a.** If the pH = 2.0 for a solution, calculate the pOH.

9.27 **b.** If the pOH = 4.73 for a solution, calculate the pH.

3.0 **c.** If the $[H_3O^+] = 1 \times 10^{-3}$ M for a solution, calculate the pH.

1×10^{-5} M **d.** If the pOH = 5.0 for a solution, calculate the $[OH^-]$.

1×10^{-13} M **e.** If the pH = 1.0 for a solution, calculate the $[OH^-]$.

3. Calculate the following values:

4.631 **a.** If the $[H_3O^+] = 2.34 \times 10^{-5}$ M for a solution, calculate the pH.

3×10^{-4} M **b.** If the pOH = 3.5 for a solution, calculate the $[OH^-]$.

2.2×10^{-7} M **c.** If the $[H_3O^+] = 4.6 \times 10^{-8}$ M for a solution, calculate the $[OH^-]$.

PROBLEMS Write the answer on the line to the left. Show all your work in the space provided.

4. The $[H_3O^+] = 2.3 \times 10^{-3}$ M for an aqueous solution.

4.3×10^{-12} M **a.** Calculate $[OH^-]$ in this solution.

2.64 b. Calculate the pH of this solution.

11.36 c. Calculate the pOH of this solution.

d. Is the solution acidic, basic, or neutral? Explain your answer.

acidic because the pH is less than 7.0

5. Consider a dilute solution of 0.025 M $Ba(OH)_2$ to answer the following questions.

a. What is the $[OH^-]$ of this solution? Explain your answer.

0.050 M; In solution, $Ba(OH)_2$ releases two OH^- per molecule, so doubling the $[Ba(OH)_2]$ equals the $[OH^-]$.

12.70 b. What is the pH of this solution?

6. Vinegar purchased in a store may contain 6 g of CH_3COOH per 100 mL of solution.

1 M a. What is the molarity of the solute?

b. The actual $[H_3O^+]$ of the vinegar solution in part a is 4.2×10^{-3} M. In this solution, has more than 1% or less than 1% of the acetic acid ionized? Explain your answer.

less than 1%; 1% of 1 M would equal 1×10^{-2} hydronium ions, but in fact less than that amount is produced.

weak c. Is acetic acid strong or weak, based on the ionization information from part b?

2.38 d. What is the pH of this vinegar solution?

Name ______________________ Date ______________ Class ______________

CHAPTER 16 REVIEW

Acid-Base Titration and pH

SECTION 16-2

SHORT ANSWER **Answer the following questions in the space provided.**

1. Below is a pH curve from an acid-base titration. On it are labeled three points: X, Y, and Z.

Y **a.** Which point represents the equivalence point?

Z **b.** At which point is there excess acid in the system?

X **c.** At which point is there excess base in the system?

0.00750 mol **d.** If the base solution is 0.250 M, how many moles of OH^- are consumed at the end point of the titration?

Acid-Base Titration Curve

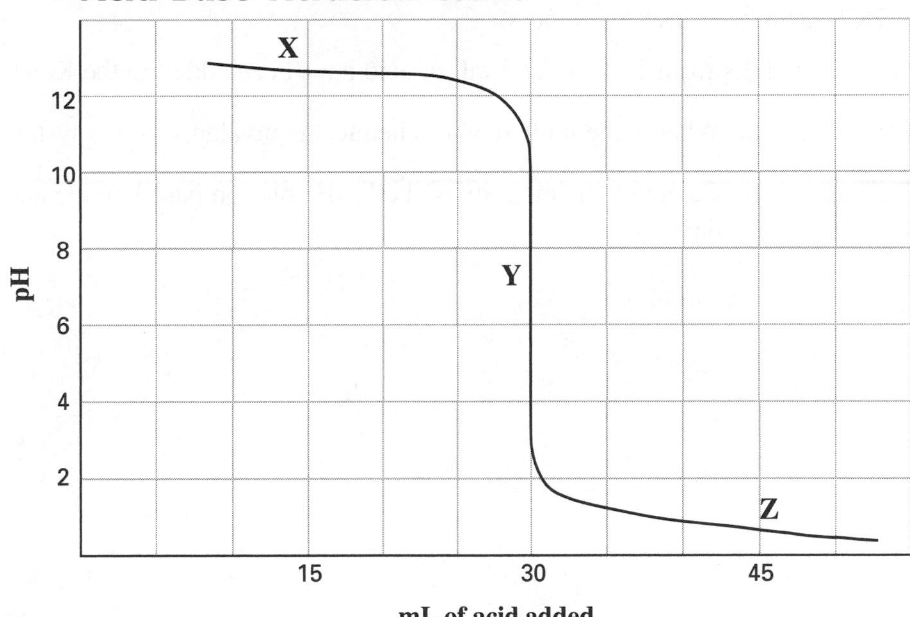

PROBLEMS **Write the answer on the line to the left. Show all your work in the space provided.**

2. A standardized solution of 0.065 M HCl is titrated with a saturated solution of calcium hydroxide to determine its molarity and its solubility. It takes 25.0 mL of base to neutralize 10.0 mL of the acid.

a. Write the balanced molecular equation for this neutralization reaction.

$Ca(OH)_2(aq) + 2HCl(aq) \rightarrow CaCl_2(aq) + 2H_2O(l)$

SECTION 16-2 continued

___0.013 M___ **b.** Determine the molarity of the $Ca(OH)_2$ solution.

___0.96 g/L___ **c.** Based on your answer to part b, calculate the solubility of the base in grams per liter of solution.

3. It is possible to carry out a titration without any indicator present. Instead, a pH probe is immersed in a beaker containing the solution of unknown molarity. The solution of known molarity is slowly added from a buret. Use the titration data below to answer the following questions:

Volume of KOH(*aq*) in the beaker = 30.0 mL

Molarity of HCl(*aq*) in the burette = 0.50 M

At the instant the pH falls from 10 to 4, 27.8 mL of acid have been added to the KOH in the beaker.

___1:1___ **a.** What is the mole ratio of chemical equivalents in this system?

___0.46 M___ **b.** Calculate the molarity of the KOH solution based on the above data.

Name ______________________ Date ______________ Class ______________

CHAPTER 16 REVIEW

Acid-Base Titration and pH

MIXED REVIEW

SHORT ANSWER **Answer the following questions in the space provided.**

1. Calculate the following values: (A calculator should not be necessary.)

4.0 ______ **a.** If the $[H_3O^+] = 1 \times 10^{-4}$ M for a solution, calculate the pH.

1×10^{-13} M ______ **b.** If the pH = 13.0 for a solution, calculate the $[H_3O^+]$.

1×10^{-9} M ______ **c.** If the $[OH^-] = 1 \times 10^{-5}$ M for a solution, calculate the $[H_3O^+]$.

9.28 ______ **d.** If the pH = 4.72 for a solution, calculate the pOH.

14.00 ______ **e.** If the $[OH^-] = 1.0$ M for a solution, calculate the pH.

2. Calculate the following values:

8.204 ______ **a.** If the $[H_3O^+] = 6.25 \times 10^{-9}$ M for a solution, calculate the pH.

4.6×10^{-3} M ______ **b.** If the pOH = 2.34 for a solution, calculate the $[OH^-]$.

3×10^{-4} M ______ **c.** The $[OH^-]$ in milk of magnesia is ____. (Use data from Table 16-4 on page 486 of the text.)

PROBLEMS **Write the answer on the line to the left. Show all your work in the space provided.**

3. A 0.0012 M solution of H_2SO_4 is 100% ionized.

0.0024 M ______ **a.** What is the $[H_3O^+]$ of the H_2SO_4 solution?

4.2×10^{-12} M ______ **b.** What is the $[OH^-]$ of this solution?

2.62 ______ **c.** What is the pH of this solution?

MIXED REVIEW continued

4. In a titration, a 25.0 mL sample of 0.150 M HCl is neutralized with 44.45 mL of $Ba(OH)_2$.

a. Write the balanced molecular equation for this reaction.

$2HCl(aq) + Ba(OH)_2(aq) \rightarrow BaCl_2(aq) + 2H_2O(l)$

0.0422 M **b.** What is the molarity of the base solution?

5. 3.09 g of boric acid, H_3BO_3, are dissolved into 200 mL of solution.

0.2 M **a.** Calculate the molarity of the solution.

b. H_3BO_3 ionizes in solution in three stages. Write the equation showing the ionization for each stage. Which stage proceeds furthest to completion?

stage 1: $H_3BO_3(s) + H_2O(l) \rightleftarrows H_3O^+(aq) + H_2BO_3^-(aq)$

stage 2: $H_2BO_3^-(aq) + H_2O(l) \rightleftarrows H_3O^+(aq) + HBO_3^{2-}(aq)$

stage 3: $HBO_3^{2-}(aq) + H_2O(l) \rightleftarrows H_3O^+(aq) + BO_3^{3-}(aq)$

Stage 1 proceeds furthest to completion.

1.3×10^{-5} M **c.** What is the $[H_3O^+]$ for this boric acid solution if the pH = 4.90?

less than 1% **d.** Is the percent ionization of this H_3BO_3 solution more than or less than 1%?

Name ____________________ Date ____________ Class ____________

CHAPTER 17 REVIEW

Reaction Energy and Reaction Kinetics

SECTION 17-1

SHORT ANSWER Answer the following questions in the space provided.

1. ___0___ For elements in their standard state, the value of ΔH_f^o is ____.

2. The formation and decomposition of water can be represented by the following thermochemical equations:

$$H_2(g) + \frac{1}{2}O_2(g) \rightarrow H_2O(g) + 241.8 \text{ kJ/mol}$$

$$H_2O(l) + 241.8 \text{ kJ/mol} \rightarrow H_2(g) + \frac{1}{2}O_2(g)$$

___taken in___ **a.** Is heat being taken in or released as liquid H_2O decomposes?

___positive___ **b.** What is the appropriate sign for the enthalpy change in this decomposition reaction?

PROBLEMS Write the answer on the line to the left. Show all your work in the space provided.

3. ___70°C___ If 200. g of water at 20°C absorbs 41 840 J of heat, what will its final temperature be?

4. ___28.9 kJ___ Aluminum has a specific heat of 0.900 J/(g•°C). How much energy in kJ is needed to raise the temperature of a 625 g block of aluminum from 30.7°C to 82.1°C?

5. The products in a reaction have a total heat content of 458 kJ/mol and the reactants have a total heat content of 658 kJ/mol.

___−200. kJ/mol___ **a.** What is the value of ΔH for this reaction?

products _______ **b.** Which is the more stable part of this system, the reactants or the products?

6. The heat of combustion of acetylene gas is −1301.1 kJ/mol of C_2H_2.

a. Write the balanced thermochemical equation for the complete combustion of C_2H_2.

$C_2H_2(g) + \frac{5}{2}O_2(g) \rightarrow 2CO_2(g) + H_2O(l)$ + heat energy

320 kJ _______ **b.** If 0.25 mol of C_2H_2 react according to the equation in part a, how much heat is released?

78 g _______ **c.** How many grams of C_2H_2 are needed to react according to the equation in part a to release 3900 kJ of heat?

7. 8041.9 kJ/mol _______ When 1 mol of Al_2O_3 is formed according to the equation below, 1169.8 kJ of heat are liberated. Determine the ΔH for the reaction between Al and Fe_3O_4 if the heat of formation for Fe_3O_4 is −1120.9 kJ/mol.

$$8Al(s) + 3Fe_3O_4(s) \rightarrow 4Al_2O_3(s) + 9Fe(s)$$

8. −196.0 kJ/mol _______ Use the data in Appendix Table A-14 of the text to determine the ΔH of the following reaction.

$$2H_2O_2(l) \rightarrow 2H_2O(l) + O_2(g)$$

Name ______________________ Date ______________ Class ______________

CHAPTER 17 REVIEW

Reaction Energy and Reaction Kinetics

SECTION 17-2

SHORT ANSWER **Answer the following questions in the space provided.**

1. For the following examples, state whether the change in entropy favors the forward or reverse reaction:

forward reaction **a.** $HCl(l) \rightleftarrows HCl(g)$

reverse reaction **b.** $C_6H_{12}O_6(aq) \rightleftarrows C_6H_{12}O_6(s)$

forward reaction **c.** $2NH_3(g) \rightleftarrows N_2(g) + 3H_2(g)$

reverse reaction **d.** $3C_2H_4(g) \rightleftarrows C_6H_{12}(l)$

2. $\Delta G = \Delta H - T\Delta S$ **a.** Write an equation that shows the relationship between enthalpy, entropy, and free energy.

negative **b.** For a reaction to occur spontaneously, the sign of ΔG should be ____.

3. Consider the following reaction: $NH_3(g) + H_2O(l) \rightleftarrows NH_4^+(aq) + OH^-(aq)$ + heat energy

True **a.** The enthalpy factor favors the forward reaction. True or False?

reverse reaction **b.** The sign of $T\Delta S^o$ is negative. This means the entropy factor favors the ____.

c. Given that ΔG^o for the above reaction is positive, which term is greater in magnitude and therefore predominates, $T\Delta S$ or ΔH?

$T\Delta S > \Delta H$; ΔH is negative, but ΔG is positive, indicating the reaction is not spontaneous. Entropy must be stronger than enthalpy in this system.

4. Consider the following equation for the vaporization of water:

$$H_2O(l) \rightleftarrows H_2O(g) \qquad \Delta H = +40.65 \text{ kJ/mol at } 100°C$$

endothermic **a.** Is the forward reaction exothermic or endothermic?

the reverse reaction **b.** Does the enthalpy factor favor the forward or reverse reaction?

the forward reaction **c.** Does the entropy factor favor the forward or reverse reaction?

Name ______________________ Date ______________ Class ______________

SECTION 17-2 continued

PROBLEMS **Write the answer on the line to the left. Show all your work in the space provided.**

5. Halogens can combine with other halogens to form several unstable compounds. Consider the following equation: $I_2(s) + Cl_2(g) \rightleftarrows 2ICl(g)$
ΔH_f^o for the formation of ICl = +18.0 kJ/mol and ΔG^o = −5.4 kJ/mol.

the reverse reaction ______ **a.** Is the forward or reverse reaction favored by the enthalpy factor?

the forward reaction ______ **b.** Will the forward or reverse reaction occur spontaneously at standard conditions?

the forward reaction ______ **c.** Is the forward or reverse reaction favored by the entropy factor?

+23.4 kJ/(mol·K) ______ **d.** Calculate the value of $T\Delta S$ for this system.

0.0785 kJ/(mol·K) ______ **e.** Calculate the value of ΔS for this system at 25°C.

6. Calculate the free energy change for the reactions below. Determine whether each reaction will be spontaneous or nonspontaneous.

−51.0 kJ/mol; spontaneous ______ **a.** $C(s) + 2H_2(g) \rightarrow CH_4(g)$

$\Delta S^O = -80.7$ J/(mol•K), $\Delta H^O = -75.0$ kJ/mol,
$T = 298$ K

195.8 kJ/mol; nonspontaneous ______ **b.** $3Fe_2O_3(s) \rightarrow 2Fe_3O_4(s) + \frac{1}{2}O_2(g)$

$\Delta S^O = 134.2$ J/(mol•K), $\Delta H^O = 235.8$ kJ/mol,
$T = 298$ K

Name ______________________ Date ______________ Class ______________

CHAPTER 17 REVIEW

Reaction Energy and Reaction Kinetics

SECTION 17-3

SHORT ANSWER **Answer the following questions in the space provided.**

1. Refer to the energy diagram at the bottom of this page to answer the following questions:

__d__ **a.** Which letter represents the energy of the activated complex?

(a) A (c) C
(b) B (d) D

__c__ **b.** Which letter represents the energy of the reactants?

(a) A (c) C
(b) B (d) D

__d__ **c.** Which of the following choices represents the quantity of activation energy for the forward reaction?

(a) the amount of energy at C minus the amount of energy at B
(b) the amount of energy at D minus the amount of energy at A
(c) the amount of energy at D minus the amount of energy at B
(d) the amount of energy at D minus the amount of energy at C

__c__ **d.** Which of the following choices represents the quantity of activation energy for the reverse reaction?

(a) the amount of energy at C minus the amount of energy at B
(b) the amount of energy at D minus the amount of energy at A
(c) the amount of energy at D minus the amount of energy at B
(d) the amount of energy at D minus the amount of energy at C

__b__ **e.** Which of the following choices represents the quantity of the heat of reaction for the forward reaction?

(a) the amount of energy at C minus the amount of energy at B
(b) the amount of energy at B minus the amount of energy at C
(c) the amount of energy at D minus the amount of energy at B
(d) the amount of energy at B minus the amount of energy at A

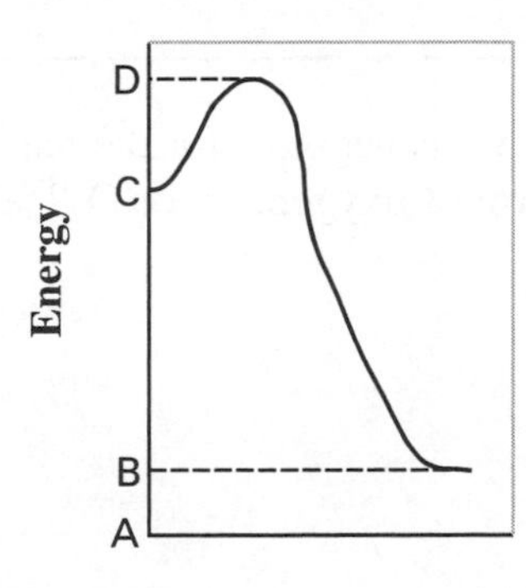

SECTION 17-3 continued

2. For the reaction A + B → X, the activation energy for the forward direction equals 85 kJ/mol and the activation energy for the reverse direction equals 80 kJ/mol.

______the product______ **a.** Which side has the greater energy content, the reactants or the product?

______+5 kJ/mol______ **b.** What is the heat of reaction in the forward direction?

______True______ **c.** The heat of reaction in the reverse direction is equal in magnitude but opposite in sign to the heat of reaction in the forward direction. True or False?

3. Below is an incomplete energy diagram.

a. Use the following data to complete the diagram: $E_a = +50$ kJ/mol; $\Delta E_{forward} = -10$ kJ/mol. Label the reactants, products, ΔE, E_a, E_a', and the activated complex.

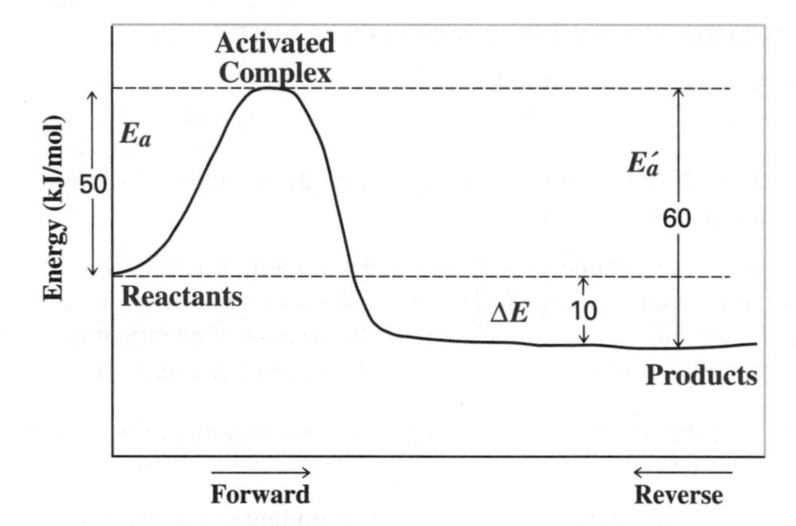

______+60 kJ/mol______ **b.** What is the value of E_a'?

4. It is proposed that ozone undergoes the following two-step mechanism in our upper atmosphere:

$$O_3(g) \rightarrow O_2(g) + O(g)$$
$$O_3(g) + O(g) \rightarrow 2O_2(g)$$

a. Identify any intermediates formed in the above equations.

Monatomic O is the intermediate formed.

b. Write the net equation.

$2O_3(g) \rightarrow 3O_2(g)$

______O_2______ **c.** If ΔH is negative for the reaction in part b, which is the more stable form of oxygen, O_3 or O_2?

Name ________________________ Date ______________ Class ______________

CHAPTER 17 REVIEW

Reaction Energy and Reaction Kinetics

SECTION 17-4

SHORT ANSWER **Answer the following questions in the space provided.**

1. Below is an energy diagram for a particular process. One curve represents the energy profile for the uncatalyzed reaction, and the other curve represents the energy profile for the catalyzed reaction.

___a___ **a.** Which curve has the greater activation energy?

(a) curve 1
(b) curve 2
(c) Both are equal.

___c___ **b.** Which curve has the greater heat of reaction?

(a) curve 1
(b) curve 2
(c) Both are equal.

___b___ **c.** Which curve represents the catalyzed process?

(a) curve 1
(b) curve 2

d. Explain your answer to part c.

The catalyst has formed an alternative activated complex that requires a lower activation energy, as represented by the lower curve.

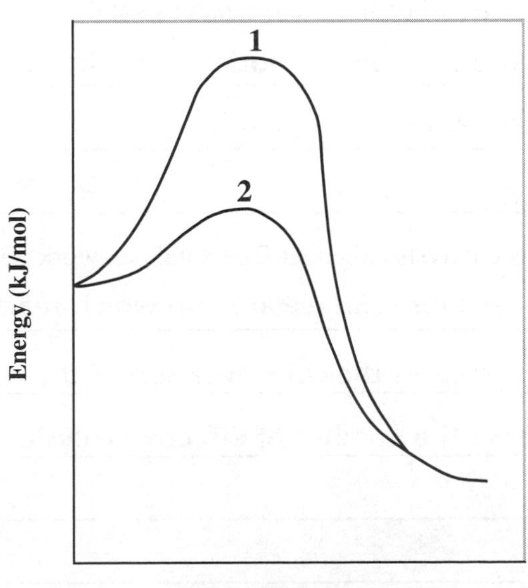

2. Is it correct to say that a catalyst affects the speed of a reaction but does not take part in the reaction? Explain your answer.

It is not correct. The catalyst does take part in the reaction. However, if it is used up in one step of the mechanism, it is regenerated in a later step. There is no net change in mass for the catalyst.

3. The reaction X + Y → Z is shown to have the following rate law:

$$R = k[X]^3[Y]$$

a. What is the effect on the rate if the concentration of Y is reduced by one-third?

The rate is reduced by one-third as well.

b. What is the effect on the rate if the concentration of X is doubled?

The rate increases by a factor of eight.

c. What is the effect on the rate if a catalyst is added to the system?

The rate will increase if the catalyst added is specific for this reaction.

4. Explain the following statements using collision theory:

a. Gaseous reactants react faster under high pressure than under low pressure.

At high pressure, gas molecules are more closely packed and collide more frequently. Thus, more-effective collisions occur per unit of time.

b. Ionic compounds react faster when in solution than as solids.

Ions in solution have more freedom of motion than do ions in a solid; therefore, they can collide with one another more frequently.

c. A class of heterogeneous catalysts called surface catalysts work best as a fine powder.

The fine powder produces more surface area on which reactant particles can be absorbed and in effect increases the concentration of the reactants. An increase in concentration will increase the number of effective collisions between reactant particles.

Name ______________________ Date ______________ Class ______________

CHAPTER 17 REVIEW

Reaction Energy and Reaction Kinetics

MIXED REVIEW

SHORT ANSWER **Answer the following questions in the space provided.**

1. Given the following reaction for the decomposition of hydrogen peroxide:

$$2H_2O_2(l) \rightarrow 2H_2O(l) + O_2(g)$$

List three ways to speed up the rate of decomposition. For each one, briefly explain why it is effective based on collision theory.

increase the concentration of hydrogen peroxide—allows more collisions per unit of time to occur

increase the temperature of the solution—allows more energetic collisions per unit of time to occur

stir the solution—exposes more reactant surface area for the collisions to occur

add a catalyst—lowers the activation energy so that more-effective collisions can occur

2. The following equation represents a reaction that is strongly favored in the forward direction:

$$2C_7H_5(NO_2)_3(l) + 12O_2(g) \rightarrow 14CO_2(g) + 5H_2O(g) + 3N_2O(g) + \text{heat}$$

a. Why would ΔG be negative in the above reaction?

Both energy and entropy factors favor the forward spontaneous reaction. The reaction is exothermic, and there are more gas molecules in the products than in the reactants.

b. The above reaction does not occur immediately when the reactants make contact. What does this imply about the necessary activation energy?

There must be a large E_a to be overcome.

3. An ingredient in smog is the gas NO. One reaction that controls the concentration of NO in the air follows:

$$H_2(g) + 2NO(g) \rightarrow H_2O(g) + N_2O(g)$$

At high temperatures, doubling the concentration of H_2 doubles the rate of reaction, while doubling the concentration of NO increases the rate fourfold.

Write a rate law for this reaction consistent with this data.

$R = k[H_2][NO]^2$

MIXED REVIEW continued

PROBLEMS **Write the answer on the line to the left. Show all your work in the space provided.**

4. Consider the following equation and data: $2NO_2(g) \rightarrow N_2O_4(g)$

$$\Delta H_f^0 \text{ of } N_2O_4 = +9.2 \text{ kJ/mol}$$

$$\Delta H_f^0 \text{ of } NO_2 = +33.2 \text{ kJ/mol}$$

$$\Delta G^0 = -4.7 \text{ kJ/mol } N_2O_4$$

Use Hess's law to calculate ΔH^0 for the above reaction.

$N_2(g) + 2O_2(g) \rightarrow N_2O_4(g)$ **$\Delta H_f^0 = +9.2$ kJ/mol**

$2NO_2(g) \rightarrow N_2(g) + 2O_2(g)$ **$\Delta H_f^0 = 2(-33.2$ kJ/mol)**

$2NO_2(g) \rightarrow N_2O_4(g)$ **$\Delta H^0 = -57.2$ kJ/mol**

5. Answer the following questions using the energy diagram at the bottom of the page.

endothermic **a.** Is the reaction represented by the curve exothermic or endothermic?

+40 kJ/mol **b.** Estimate the magnitude and sign of $\Delta E_{forward}$.

+20 kJ/mol **c.** Estimate E_a'.

A catalyst is added to the reaction that lowers E_a by about 15 kJ/mol.

speed up **d.** Does the forward reaction rate speed up or slow down?

speed up **e.** Does the reverse reaction rate speed up or slow down?

No **f.** Does $\Delta E_{forward}$ change from its value in part b?

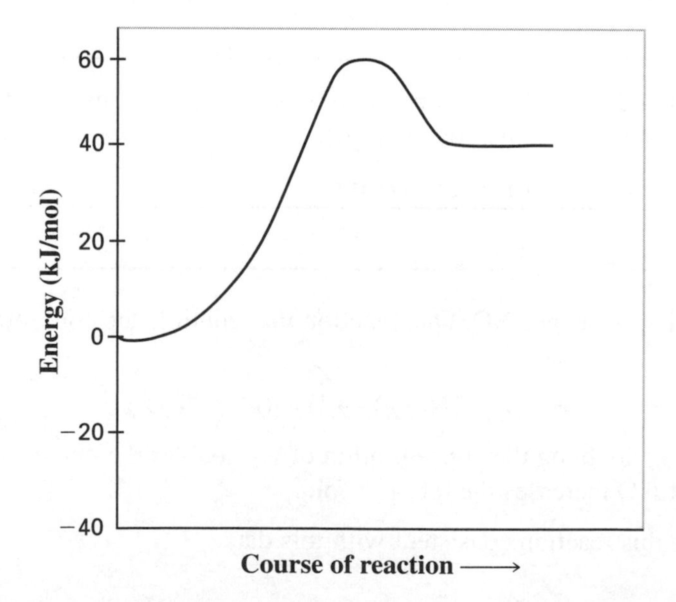

Name ______________________ Date ______________ Class ______________

CHAPTER 18 REVIEW

Chemical Equilibrium

SECTION 18-1

SHORT ANSWER **Answer the following questions in the space provided.**

1. __c__ Silver chromate dissolves in water according to the following equation:

$$Ag_2CrO_4(s) \rightleftarrows 2Ag^+(aq) + CrO_4^{2-}(aq)$$

Which of these correctly represents the equilibrium expression for the above equation?

(a) $\frac{2[Ag^+] + [CrO_4^{2-}]}{Ag_2CrO_4}$ (b) $\frac{[Ag_2CrO_4]}{[Ag^+]^2[CrO_4^{2-}]}$ (c) $\frac{[Ag^+]^2[CrO_4^{2-}]}{1}$ (d) $\frac{[Ag^+]^2[CrO_4^{2-}]}{2[Ag_2CrO_4]}$

2. Are pure solids included in equilibrium expressions? Explain your answer.

Pure solids are not included in equilibrium expressions because their concentrations do not change. Their constant value is incorporated into *K*.

3. Write the equilibrium expression for the following hypothetical equation:

$$3A(aq) + B(aq) \rightleftarrows 2C(aq) + 3D(aq)$$

$$K = \frac{[C]^2[D]^3}{[A]^3[B]}$$

4. a. Write the appropriate equilibrium expression for each of the following equations. Include the value of *K*.

(1) $N_2O_4(g) \rightleftarrows 2NO_2(g)$ $K = 0.1$

$$\frac{[NO]^2}{[N_2O_4]} = 0.1$$

(2) $NH_4OH(aq) \rightleftarrows NH_4^+(aq) + OH^-(aq)$ $K = 2 \times 10^{-5}$

$$\frac{[NH_4^+][OH^-]}{[NH_4OH]} = 2 \times 10^{-5}$$

SECTION 18-1 continued

(3) $PbI_2(s) \rightleftarrows Pb^{2+}(aq) + 2I^-(aq)$ $K = 7 \times 10^{-9}$

$$\frac{[Pb^{2+}][I^-]^2}{1} = 7 \times 10^{-9}$$

(4) $H_3O^+(aq) + OH^-(aq) \rightleftarrows 2H_2O(l)$ $K = 1 \times 10^{14}$

$$\frac{1}{[H_3O^+][OH^-]} = 1 \times 10^{14}$$

system 4 **b.** Which of the four systems in part a proceeds farthest to the right when equilibrium is established?

system 3 **c.** Which system contains mostly reactants at equilibrium?

PROBLEMS Write the answer on the line to the left. Show all your work in the space provided.

5. 1.1 Consider the following reaction:

$$2NO(g) + O_2(g) \rightleftarrows 2NO_2(g)$$

At equilibrium, [NO] = 0.80 M, $[O_2]$ = 0.50 M, and $[NO_2]$ = 0.60 M. Calculate the value of K for this reaction.

6. Nitrous acid is a weak acid that hydrolyzes according to the following equation:

$$HNO_2(aq) + H_2O(l) \rightleftarrows H_3O^+(aq) + NO_2^-(aq)$$

At 298 K, $K = 4.3 \times 10^{-4}$.

H_2O **a.** Which term in the above equation does not appear in the equilibrium expression?

4.3 M **b.** For the above reaction, $[H_3O^+] = [NO_2^-]$ = 0.043 M at equilibrium. Calculate $[HNO_2]$.

Name ______________________ Date ______________ Class ______________

CHAPTER 18 REVIEW

Chemical Equilibrium

SECTION 18-2

SHORT ANSWER **Answer the following questions in the space provided.**

1. __d__ Raising the temperature of any equilibrium system always results in ____.

(a) the equilibrium shifting to the left
(b) the equilibrium shifting to the right
(c) the equilibrium shifting in the direction of the exothermic reaction
(d) the equilibrium shifting in the direction of the endothermic reaction
(e) the equilibrium shifting in the direction that produces fewer molecules in the system

2. Given the equilibrium equation: $CH_3OH(g) + 101\ kJ \rightleftarrows CO(g) + 2H_2(g)$,

__b__ **a.** increasing [CO] will cause $[H_2]$ to ____.

(a) increase (b) decrease (c) not change

__b__ **b.** raising the temperature will cause the equilibrium of the system to shift ____.

(a) to the left (b) to the right (c) in neither direction

__a__ **c.** raising the temperature will cause the value of K to ____.

(a) increase (b) decrease (c) not change

3. Consider the following equilibrium equation: $H_2O(g) + C(s) \rightleftarrows H_2(g) + CO(g) +$ heat energy
Which way will the equilibrium of the system shift (left, right, or in neither direction) when

__left__ **a.** extra CO gas is introduced?

__neither direction__ **b.** a catalyst is introduced?

__right__ **c.** the temperature of the system is lowered?

__left__ **d.** the pressure on the system is increased by decreasing the container volume?

4. Adding a catalyst to a reaction does speed up the rate of that reaction. Explain why the concentration of D does not change when a catalyst is added to the reaction represented by the following general equilibrium equation: $A + B \rightleftarrows D$

A catalyst increases the rates of the forward and reverse reactions in the system by equal factors, so D is being used up as quickly as it is produced.

5. A key step in manufacturing sulfuric acid is represented by the following equation:

$$2SO_2(g) + O_2(g) \rightleftarrows 2SO_3(g) + 100 \text{ kJ/mol}$$

To be economically viable, this process must yield as much SO_3 as possible in the shortest amount of time. You are in charge of this manufacturing process.

a. Would you impose a high pressure or a low pressure on the system? Explain your answer.

Impose a high pressure and the reaction that will produce fewer gas molecules is favored, which in this case is the forward reaction.

b. To maximize the yield of SO_3, should the temperature be kept high or low during the reaction?

A low temperature favors the forward exothermic reaction.

c. Will adding a catalyst change the yield of SO_3?

No, a catalyst does not change the percent yield.

6. The equation for an equilibrium system easily studied in a lab follows:

$$2NO_2(g) \rightleftarrows N_2O_4(g)$$

N_2O_4 gas is colorless, and NO_2 gas is dark brown. Lowering the temperature of the equilibrium mixture of gases reduces the intensity of the color.

a. Is the forward or reverse reaction favored when the temperature is lowered?

The forward reaction is favored. As the color becomes less intense, the equilibrium shifts in the direction that produces the colorless gas, N_2O_4.

b. Will the sign of ΔH be positive or negative if the temperature is lowered? Explain your answer.

Negative; lowering the temperature favors the exothermic reaction, releasing heat that keeps the temperature of the system from rising. The equilibrium shifts to the right. Therefore, the forward reaction is exothermic, and ΔH has a negative sign.

Name ______________________ Date ______________ Class ______________

CHAPTER 18 REVIEW

Chemical Equilibrium

SECTION 18-3

SHORT ANSWER **Answer the following questions in the space provided.**

1. __a__ The pH of tomato juice is 4.3, indicating that tomato juice is ____.

(a) acidic
(b) basic
(c) neutral

2. __d__ Addition of the salt of a weak acid to a solution of the weak acid ____.

(a) lowers the concentration of the nonionized acid and the concentration of the H_3O^+
(b) lowers the concentration of the nonionized acid and raises the concentration of the H_3O^+
(c) raises the concentration of the nonionized acid and the concentration of the H_3O^+
(d) raises the concentration of the nonionized acid and lowers the concentration of the H_3O^+

3. __b__ Salts of a weak acid and a strong base produce solutions that are ____.

(a) acidic only
(b) basic only
(c) neutral only
(d) either acidic, basic, or neutral

4. __a__ If an acid is added to a solution of a weak base and its salt, ____.

(a) more water is formed and more weak base ionizes
(b) hydronium-ion concentration decreases
(c) more hydroxide ion is formed
(d) more nonionized weak base is formed

5. a. In the space below each of the following equations, correctly label the two conjugate acid-base pairs as *acid 1, acid 2, base 1,* and *base 2.*

(a) $CO_3^{2-}(aq) + H_3O^+(aq) \rightleftarrows HCO_3^-(aq) + H_2O(l)$

base 1 **acid 2** **acid 1** **base 2**

(b) $HPO_4^{2-}(aq) + H_2O(l) \rightleftarrows OH^-(aq) + H_2PO_4^-(aq)$

base 1 **acid 2** **base 2** **acid 1**

__b__ **b.** Which reaction in part a is an example of hydrolysis?

__a__ **c.** As the first reaction in part a proceeds, the pH of the solution ____.

(a) increases (b) decreases (c) stays at the same level

6. Write the formulas for the acid and the base that could form the salt $Ca(NO_3)_2$.

The acid is $HNO_3(aq)$ and the base is $Ca(OH)_2(aq)$.

PROBLEMS **Write the answer on the line to the left. Show all your work in the space provided.**

7. An unknown acid X hydrolyzes according to the equation in part a below.

a. In the space below the equation, correctly label the two conjugate acid-base pairs in this system as *acid 1, acid 2, base 1,* and *base 2.*

$$HX(aq) + H_2O(l) \rightleftarrows X^-(aq) + H_3O^+(aq)$$

acid 1 base 2 base 1 acid 2

b. Write the equilibrium expression for K_a for this system.

$$K_a = \frac{[X^-][H_3O^+]}{[HX]}$$

1.0×10^{-8} **c.** Experiments show that at equilibrium $[H_3O^+] = [X^-] = 2.0 \times 10^{-5}$ mol/L and $[HX] = 4.0 \times 10^{-2}$ mol/L. Calculate the value of K_a based on these data.

8. Given the following equation for the reaction of a weak base in water:

$$NH_3(aq) + H_2O(l) \rightleftarrows NH_4^+(aq) + OH^-(aq)$$

Write the equilibrium expression for K_b.

$$K_b = \frac{[NH_4^+][OH^-]}{[NH_3]}$$

Name ______________________ Date ______________ Class ______________

CHAPTER 18 REVIEW

Chemical Equilibrium

SECTION 18-4

SHORT ANSWER **Answer the following questions in the space provided.**

1. Match the solution type on the right to the corresponding relationship between the ion product and the K_{sp} for that solution listed on the left.

__c__ the ion product exceeds the K_{sp}

__a__ the ion product equals the K_{sp}

__b__ the ion product is less than the K_{sp}

(a) The solution is saturated; no more solid dissolves.

(b) The solution is unsaturated; no solid is present.

(c) The solution is supersaturated and will readily precipitate.

2. Silver carbonate, Ag_2CO_3, makes a saturated solution with $K_{sp} = 10^{-11}$.

a. Write the equilibrium expression for the dissolution of Ag_2CO_3.

$K_{sp} = [Ag^+]^2[CO_3^{2-}]$

____reverse reaction____ **b.** In this system, will the foward or reverse reaction be favored if extra Ag^+ is added?

PROBLEMS **Write the answer on the line to the left. Show all your work in the space provided.**

3. When the ionic solid XCl_2 dissolves in pure water to make a saturated solution, experiments show that 2×10^{-3} mol/L of X^{2+} ions go into solution.

a. Write the equation showing the dissolution of XCl_2 and the corresponding equilibrium expression.

$XCl_2(s) \rightleftarrows X^{2+}(aq) + 2Cl^-(aq)$

$K_{sp} = [X^{2+}][Cl^-]^2$

____3×10^{-8}____ **b.** Calculate the value of K_{sp} for XCl_2.

SECTION 18-4 continued

less soluble ______ **c.** Refer to Table 18-3 on page 579 of the text. Would XCl_2 be more soluble or less soluble than $PbCl_2$ at the same temperature?

4. The solubility of Ag_3PO_4 is 2.1×10^{-4} g/100. g.

a. Write the equation showing the dissolution of this ionic solid.

$Ag_3PO_4(s) \rightleftarrows 3Ag^+(aq) + PO_4^{3-}(aq)$

5.0×10^{-6} M ______ **b.** Calculate the molarity of this saturated solution.

1.7×10^{-20} ______ **c.** What is the value of K_{sp} for this system?

5. As $PbCl_2$ dissolves, $[Pb^{2+}] = 2.0 \times 10^{-1}$ mol/L and $[Cl^-] = 1.5 \times 10^{-2}$ mol/L.

a. Write the equilibrium expression for the dissolution of $PbCl_2$.

$K_{sp} = [Pb^{2+}][Cl^-]^2$

4.5×10^{-5} ______ **b.** Compute the ion product using the data given above.

Name ______________________ Date ______________ Class ______________

CHAPTER 18 REVIEW

Chemical Equilibrium

MIXED REVIEW

SHORT ANSWER **Answer the following questions in the space provided.**

1. Consider the following equilibrium system:

$$N_2(g) + 2O_2(g) \rightleftarrows 2NO_2(g); \Delta H = +33 \text{ kJ/mol}$$

Which reaction is favored when

reverse reaction **a.** some N_2 is removed?

neither reaction **b.** a catalyst is introduced?

forward reaction **c.** pressure on the system is increased by decreasing the volume?

forward reaction **d.** the temperature of the system is increased?

2. Ammonia gas dissolves in water according to the following equation:

$$NH_3(g) + H_2O(l) \rightleftarrows NH_4^+(aq) + OH^-(aq) + \text{heat}; K = 1.8 \times 10^{-5}$$

base **a.** Is aqueous ammonia an acid or a base?

Yes **b.** Is the equation given above an example of hydrolysis?

reverse reaction **c.** Using the value for K does the equilibrium favor the forward or reverse reaction?

PROBLEMS **Write the answer on the line to the left. Show all your work in the space provided.**

3. Formic acid, HCOOH, is a weak acid present in the venom of red-ant bites. At equilibrium [HCOOH] = 2.00 M, $[HCOO^-] = 4.0 \times 10^{-1}$ M, and $[H_3O^+] = 9.0 = 10^{-4}$ M.

a. Write the equilibrium expression for the ionization of formic acid.

$$K_a = \frac{[H_3O^+][HCOO^-]}{[HCOOH]}$$

1.8×10^{-4} **b.** Calculate the value of K_a for this acid.

MIXED REVIEW continued

4. HF hydrolyzes according to the following equation:

$$HF(aq) + H_2O(l) \rightleftarrows H_3O^+(aq) + F^-(aq)$$

When 0.030 mol of HF dissolves in 1.0 L of water, the solution quickly ionizes to reach equilibrium. At equilibrium, the remaining [HF] = 0.027 M.

0.0030 mol/L ______ **a.** How many moles of HF ionize per liter of water to reach equilibrium?

0.0030 mol/L for both ______ **b.** What is $[F^-]$ and $[H_3O^+]$?

3.3×10^{-4} ______ **c.** What is the value of K_a for HF?

5. Refer to Table 18-3 on page 579 of the text. $CaSO_4(s)$ is only slightly soluble in water.

a. Write the equilibrium equation and equilibrium expression for the dissolution of $CaSO_4(s)$.

$CaSO_4(s) \rightleftarrows Ca^{2+}(aq) + SO_4^{2-}(aq)$
$K_{sp} = [Ca^{2+}][SO_4^{2-}] = 9.1 \times 10^{-6}$

4.1×10^{-2} g/100. g H_2O ______ **b.** Determine the solubility of $CaSO_4$ at 25°C in g/100. g H_2O.

Name ________________ Date ________ Class ________

CHAPTER 19 REVIEW

Oxidation-Reduction Reactions

SECTION 19-1

SHORT ANSWER **Answer the following questions in the space provided.**

1. __a__ All the following equations involve redox reactions *except* ____.

(a) $CaO + H_2O \rightarrow Ca(OH)_2$
(b) $2SO_2 + O_2 \rightarrow 2SO_3$
(c) $2HgO \rightarrow 2Hg + O_2$
(d) $SnCl_4 + 2FeCl_2 \rightarrow 2FeCl_3 + SnCl_2$

2. Assign the correct oxidation number to the individual atom or ion in the following:

__+4__ **a.** Mn in MnO_2

__0__ **b.** S in S_8

__−1__ **c.** Cl in $CaCl_2$

__+5__ **d.** I in IO_3^-

__+4__ **e.** C in HCO_3^-

__+3__ **f.** Fe in $Fe_2(SO_4)_3$

__+6__ **g.** S in $Fe_2(SO_4)_3$

3. In each of the following half-reactions, determine the value of x:

__8__ **a.** $S^{6+} + x\,e^- \rightarrow S^{2-}$

__1−__ **b.** $2Br^x \rightarrow Br_2 + 2e^-$

__2+__ **c.** $Sn^{4+} + 2e^- \rightarrow Sn^x$

__a, c__ **d.** Which of the above half-reactions represent reduction processes?

4. Give examples other than those listed in Table 19-1 on page 591 of the text for the following:
Answers may vary.

__NaH or CaH_2__ **a.** a compound containing H in a −1 oxidation state

__Na_2O_2 or BaO_2__ **b.** a peroxide

__SO_3^{2-} or HSO_3^-__ **c.** a polyatomic ion where S is +4

__F_2__ **d.** a substance in which F is not −1

SECTION 19-1 continued

5. OILRIG is a mnemonic device often used by students to help them understand redox reactions.

"*O*xidation *i*s *l*oss, *r*eduction *i*s *g*ain."

Explain what that phrase means—loss and gain of what?

Oxidation involves losing (or ejecting) electrons; reduction involves gaining electrons.

6. For each of the following reactions, state whether or not any oxidation and reduction is occurring, and write the oxidation-reduction half-reactions for those cases where redox does occur:

a. $Ca(OH)_2(aq) + 2HCl(aq) \rightarrow CaCl_2(aq) + 2H_2O(l)$

no redox occurring

b. $CH_4(g) + 2O_2(g) \rightarrow CO_2(g) + 2H_2O(g)$

yes; oxidation: $C^{4-} \rightarrow C^{4+} + 8e-$; reduction: $O + 2e- \rightarrow O^{2-}$

c. $2Al(s) + 3CuCl_2(aq) \rightarrow 2AlCl_3(aq) + 3Cu(s)$

yes; oxidation: $Al \rightarrow Al^{3+} + 3e-$; reduction: $Cu^{2+} + 2e- \rightarrow Cu$

7. Table 19-4 on page 615 of the text lists a half-cell reaction that represents the conversion of $Cr_2O_7^{2-}$ to Cr^{3+}.

+6 **a.** What is the oxidation number assigned to each Cr in $Cr_2O_7^{2-}$?

6e− **b.** How many electrons are needed to convert $2Cr^{3+}$ to $Cr_2O_7^{2-}$?

c. Show that this half-cell reaction balances by the number of Cr, O, and H atoms present in the products and reactants.

2 Cr atoms, 7 O atoms, 14 H atoms are present in the reactants and in the products.

Name ______________________ Date ______________ Class ______________

CHAPTER 19 REVIEW

Oxidation-Reduction Reactions

SECTION 19-2

SHORT ANSWER **Answer the following questions in the space provided.**

1. __c__ All of the following should be done in the process of balancing redox equations *except* ____.

(a) adjust coefficients to balance atoms
(b) adjust coefficients in electron equations to balance numbers of electrons lost and gained
(c) adjust subscripts to balance atoms
(d) write two separate electron equations

2. MnO_4^- can be reduced to MnO_2.

Mn = +7 and +4 **a.** Assign the oxidation number to Mn in these two species.

$3e-$ **b.** How many electrons are gained per Mn atom in this reduction?

9.0×10^{23} $e-$ **c.** If 0.50 mol of MnO_4^- are reduced, how many electrons are gained?

3. Bromide ions can be oxidized to form liquid bromine. Write the balanced oxidation half-reaction for the oxidation of bromide to bromine.

$2Br^- \rightarrow Br_2 + 2e^-$

4. Some bleaches contain aqueous chlorine as the active ingredient. Aqueous chlorine is made by dissolving chlorine gas in water. This form of chlorine is capable of oxidizing iron(II) ions to iron(III) ions.

a. Write the balanced ionic equation for the redox reaction between aqueous chlorine and iron(II).

$Cl_2(aq) + 2Fe^{2+}(aq) \rightarrow 2Fe^{3+}(aq) + 2Cl^-(aq)$

b. Write the two half-reactions involved. Label them as oxidation or reduction.

$Fe^{2+} \rightarrow Fe^{3+} + 1e-$; oxidation

$Cl_2 + 2e- \rightarrow 2Cl^-$; reduction

c. Show that the equation in part a is balanced by charge.

total charge on the left = 0 + +4 = +4; total charge on the right = +6 + (−2) = +4

5. Balance the following equations. Write the oxidation and reduction half-reactions involved.

a. $MnO_2(s) + HCl(aq) \rightarrow MnCl_2(aq) + Cl_2(g) + H_2O(l)$

$MnO_2(s) + 4HCl(aq) \rightarrow MnCl_2(aq) + Cl_2(g) + 2H_2O(l)$
oxidation: $2Cl^- \rightarrow Cl_2 + 2e^-$
reduction: $MnO_2 + 4H^+ + 2e^- \rightarrow Mn^{2+} + 2H_2O$

b. $S(s) + HNO_3(aq) \rightarrow SO_3(g) + H_2O(l) + NO_2(g)$

$S(s) + 6HNO_3(aq) \rightarrow SO_3(g) + 3H_2O(l) + 6NO_2(g)$
oxidation: $(S + 3H_2O \rightarrow SO_3 + 6H^+ + 6e^-) \times 1$
reduction: $(1e^- + NO_3^- + 2H^+ \rightarrow NO_2 + H_2O) \times 6$

c. $H_2C_2O_4(aq) + K_2CrO_4(aq) + HCl(aq) \rightarrow CrCl_3(aq) + KCl(aq) + H_2O(l) + CO_2(g)$

$3H_2C_2O_4(aq) + 2K_2CrO_4(aq) + 10HCl(aq) \rightarrow 2CrCl_3(aq) + 4KCl(aq) + 8H_2O(l) + 6CO_2(g)$
oxidation: $(C_2O_4^{2-} \rightarrow 2CO_2 + 2e^-) \times 3$
reduction: $(CrO_4^{2-} + 8H^+ + 3e^- \rightarrow Cr^{3+} + 4H_2O) \times 2$

Name ____________________ Date ____________ Class ____________

CHAPTER 19 REVIEW

Oxidation-Reduction Reactions

SECTION 19-3

SHORT ANSWER **Answer the following questions in the space provided.**

1. For each of the following, identify the stronger oxidizing or reducing agent:

Ca **a.** Ca or Cu as a reducing agent

Ag^+ **b.** Ag^+ or Na^+ as an oxidizing agent

Fe^{3+} **c.** Fe^{3+} or Fe^{2+} as an oxidizing agent

2. For each of the following incomplete reactions, state whether a redox reaction is likely to occur:

Yes **a.** $Mg + Sn^{2+} \rightarrow$

No **b.** $Ag + Cu^{2+} \rightarrow$

Yes **c.** $Br_2 + I^- \rightarrow$

3. Label each of the following statements about redox as True or False:

True **a.** A strong oxidizing agent is itself readily reduced.

True **b.** In autooxidation, one chemical acts as both an oxidizing agent and a reducing agent in the same process.

False **c.** The number of moles of chemical oxidized must equal the number of moles of chemical reduced.

4. Solutions of Fe^{2+} are fairly unstable in part because they can undergo autooxidation, as shown by the following unbalanced equation:

$$Fe^{2+} \rightarrow Fe^{3+} + Fe$$

a. Balance the above equation.

$3Fe^{2+} \rightarrow 2Fe^{3+} + Fe$

0.072 mol **b.** If the above reaction produces 0.036 mol of Fe, how many moles of Fe^{3+} will form?

5. Oxygen gas is a powerful oxidizing agent.

0 **a.** Assign the oxidation number to O_2.

−2 **b.** What does oxygen's oxidation number usually change to when it functions as an oxidizing agent?

c. Approximately where would you place O_2 in the list of oxidizing agents in Table 19-3 on page 603 of the text?

at the bottom right, near F_2, but above it

d. Describe the changes in oxidation states that occur in carbon and oxygen in the following combustion reaction, and identify the oxidizing and reducing agents:

$$C_6H_{12}O_6(s) + 6O_2(g) \rightarrow 6CO_2(g) + 6H_2O(l)$$

Each O atom in O_2 changes from the oxidation state of 0 to −2 in CO_2 and H_2O. Oxygen is reduced; therefore, it is the oxidizing agent. Each C atom in sugar is in the oxidation state of 0 as a reactant and +4 in CO_2 as a product. It is oxidized, and is therefore the reducing agent.

6. An example of autooxidation is the slow decomposition of aqueous chlorine, $Cl_2(aq)$, represented by the following unbalanced equation:

$$Cl_2(aq) + H_2O(l) \rightarrow ClO^-(aq) + Cl^-(aq) + H^+(aq)$$

a. Show that the oxygen and hydrogen atoms in the above reaction are not changing oxidation states.

O has an oxidation number of −2 and H has an oxidation number of +1 throughout the reaction.

b. Show the changes in the oxidation states of chlorine as this reaction proceeds.

$Cl^0 \rightarrow Cl^+$ in oxidation; $Cl^0 \rightarrow Cl^-$ in reduction

$1e^-$ **c.** In the oxidation reaction, how many electrons are transferred per Cl atom?

$1e^-$ **d.** In the reduction reaction, how many electrons are transferred per Cl atom?

e. What must be the ratio of ClO^- to Cl^- in the above reaction? Explain your answer.

1 ClO^- to 1 Cl^- to make the number of electrons given off and gained equal

f. Balance the equation for the decomposition of $Cl_2(aq)$.

$Cl_2(aq) + H_2O \rightarrow ClO^-(aq) + Cl^-(aq) + 2H^+(aq)$

Name ______________________ Date ______________ Class ______________

CHAPTER 19 REVIEW

Oxidation-Reduction Reactions

SECTION 19-4

SHORT ANSWER **Answer the following questions in the space provided.**

1. __c__ In a voltaic cell, transfer of charge through the external wires occurs by means of ____.

(a) ionization
(b) ion migration
(c) electron migration
(d) proton migration

2. __b__ The transfer of charge through the electrolyte solution occurs by means of ____.

(a) ionization
(b) ion migration
(c) electron migration
(d) proton migration

3. __c__ All the following claims about voltaic cells are true *except* ____.

(a) E^{o}_{cell} is positive.
(b) The redox reaction in the cell occurs without the addition of electrical energy.
(c) Electrical energy is converted to chemical energy.
(d) Chemical energy is converted to electrical energy.

4. Label each of the following statements as applying to a voltaic cell, an electrolytic cell, or both:

__both__ **a.** The cell reaction involves oxidation and reduction.

__voltaic cell__ **b.** The cell reaction proceeds spontaneously.

__electrolytic cell__ **c.** The cell reaction is endothermic.

__voltaic cell__ **d.** The cell reaction converts chemical energy into electrical energy.

__electrolytic cell__ **e.** The cell reaction converts electrical energy into chemical energy.

__both__ **f.** The cell contains both a cathode and an anode.

5. Use Table 19-4 on page 615 of the text to find E^{o} for the following:

__+0.56 V__ **a.** the reduction of MnO_4^{1-} to MnO_4^{2-}

__+0.74 V__ **b.** the oxidation of Cr to Cr^{3+}

__0__ **c.** the reaction within the SHE electrode

__+0.29 V__ **d.** $Cl_2 + 2Br^- \rightarrow 2Cl^- + Br_2$

PROBLEMS **Write the answer on the line to the left. Show all your work in the space provided.**

6. Below is a diagram of a voltaic cell.

a. Write the anode half-reaction.

$Cu(s) \rightarrow Cu^{2+}(aq) + 2e^-$

b. Write the cathode half-reaction.

$Ag^+(aq) + 1e^- \rightarrow Ag(s)$

c. Write the balanced cell reaction.

$Cu(s) + 2Ag^+(aq) \rightarrow Cu^{2+}(aq) + 2Ag(s)$

clockwise **d.** Do electrons within the voltaic cell travel through the voltmeter in a clockwise or counterclockwise direction?

clockwise **e.** Do anions in the beaker pass through the porous membrane in a clockwise or counterclockwise direction?

+0.46 V **f.** Calculate what the voltmeter should read when the cell is at standard state conditions. Use data from Table 19-4 on page 615 of the text.

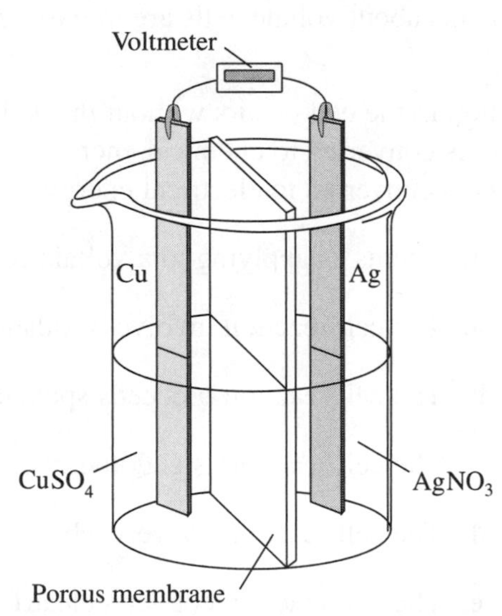

7. The silver in the voltaic cell described in item 6 is replaced with metal X and its 2+ ion. A voltage reading shows that the direction of current reverses, and the voltmeter now reads +0.74 V.

−0.40 V **a.** From these data, calculate the reduction potential of metal X.

Cd **b.** Predict the identity of metal X, based on the data in Table 19-4 on page 615 of the text.

Name ______________________ Date ______________ Class ______________

CHAPTER 19 REVIEW

Oxidation-Reduction Reactions

MIXED REVIEW

SHORT ANSWER **Answer the following questions in the space provided.**

1. Label the following descriptions of electrochemical cells as voltaic or electrolytic:

voltaic **a.** harnesses a spontaneous redox reaction to create an electric current

electrolytic **b.** uses a current from an external power supply to force a nonspontaneous redox reaction to take place

voltaic **c.** the reaction within an alkaline battery

voltaic **d.** has a positive value for E^o_{cell}

PROBLEMS **Write the answer on the line to the left. Show all your work in the space provided.**

2. Given the following unbalanced equation:

$$KMnO_4(aq) + HCl(aq) + Al(s) \rightarrow AlCl_3(aq) + MnCl_2(aq) + KCl(aq) + H_2O(l)$$

a. Write the oxidation and reduction half-reactions.

oxidation: $Al \rightarrow Al^{3+} + 3e^-$

reduction: $MnO_4^- + 8H^+ + 5e^- \rightarrow Mn^{2+} + 4H_2O$

b. Balance the equation using the seven-step procedure shown in Section 19-2 of the text.

$(MnO_4^- + 8H^+ + 5e^- \rightarrow Mn^{2+} + 4H_2O) \times 3$
$(Al \rightarrow Al^{3+} + 3e^-) \times 5$
$3KMnO_4 + 24HCl + 5Al \rightarrow 5AlCl_3 + 3MnCl_2 + 3KCl + 12H_2O$

MnO_4^- **c.** Identify the oxidizing agent in this system.

Name ______________________ Date ______________ Class ______________

MIXED REVIEW continued

+3.16 V **d.** If this reaction were to take place in an electrochemical cell, calculate the value of E^o, using the data in Table 19-4 on page 615 of the text.

e. Is the above reaction a spontaneous redox reaction? Explain your answer.

Yes, because E^o is a large positive value.

3. Given the following unbalanced ionic equation: $ClO^- + H^+ \rightarrow Cl_2 + ClO_3^- + H_2O$

a. Assign the oxidation number to each term.

in ClO^-: Cl is +1, O is −2; in H^+: H is +1;

in Cl_2: Cl is 0; in ClO_3^-: Cl is +5, O is −2; in H_2O: H is +1, O is −2

4 *e*− **b.** How many electrons are given up by each Cl atom as they are oxidized?

1 *e*− **c.** How many electrons are gained by each Cl atom as they are reduced?

Yes **d.** Is this an example of autooxidation?

e. Balance the above equation using the method of your choice.

$5ClO^- + 4H^+ \rightarrow 2Cl_2 + ClO_3^- + 2H_2O$

Name ______________________ Date ______________ Class ______________

CHAPTER 20 REVIEW

Carbon and Hydrocarbons

SECTION 20-1

SHORT ANSWER **Answer the following questions in the space provided.**

1. __d__ Carbon forms four covalent bonds that are directed in space toward the corners of a regular ____.

   (a) quadrilateral
   (b) pyramid
   (c) polygon
   (d) tetrahedron

2. __b__ The electron configuration of carbon in its ground state is ____.

   (a) $1s^22s^22p^3$
   (b) $1s^22s^22p^2$
   (c) $1s^22s^12p^2$
   (d) $1s^22s^22p^4$

3. __c__ How many covalent bonds can a carbon atom ordinarily form?

   (a) 2
   (b) 3
   (c) 4
   (d) 5

4. __c__ Carbon atoms form bonds readily with atoms of ____.

   (a) elements other than carbon
   (b) carbon only
   (c) both other elements and carbon
   (d) only neutral elements

5. __d__ The bonding between atoms in a layer of graphite consists of ____.

   (a) single bonds only
   (b) double bonds only
   (c) alternating single and double bonds
   (d) bonds that are intermediate in character between single and double bonds

6. __c__ Graphite is a good lubricant because it is arranged in layers that can slide across one another. The attractions that hold one layer to another are ____.

   (a) covalent attractions
   (b) ionic attractions
   (c) London dispersion attractions
   (d) very strong

7. __c__ The hybridization of carbon's orbitals in the CH_4 molecule is ____.

   (a) sp
   (b) sp^2
   (c) sp^3
   (d) sp^4

8. __b__ The hybridization of carbon's orbitals in the C_2H_4 molecule is ____.

   (a) sp
   (b) sp^2
   (c) sp^3
   (d) sp^4

9. __a__ The hybridization of carbon's orbitals in the C_2H_2 molecule is ____.

   (a) sp
   (b) sp^2
   (c) sp^3
   (d) sp^4

Name ____________ Date ____________ Class ____________

10. Explain why graphite conducts electricity, while diamond does not.

The delocalized electrons in graphite move freely in each layer and therefore are able to move in an electric field. The valence electrons in diamond are fixed in localized covalent bonds and cannot flow when exposed to an electric field.

11. Briefly describe the geometry of each of the allotropes of carbon.

diamond—tetrahedral

graphite—planar plates stacked together

fullerene—domelike

12. Explain why diamond conducts heat easily.

In diamonds, heat is conducted by the transfer of energy of vibration from one carbon atom to the next. The covalent bonds holding the carbon atoms together are very strong and can easily transfer vibrational energy from one atom to another.

13. Identify the allotrope of carbon represented by each of the following figures:

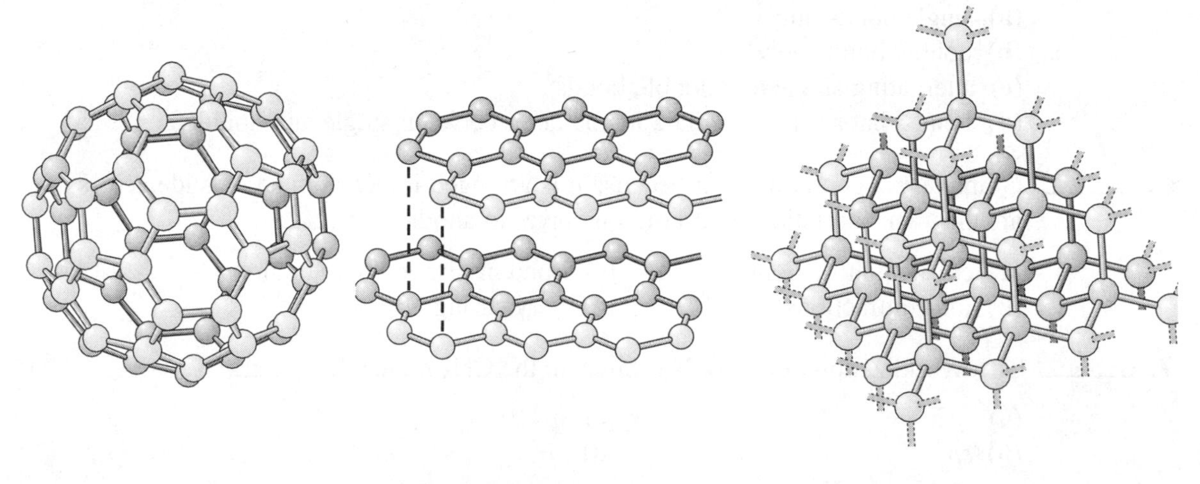

a. **fullerene** **b.** **graphite** **c.** **diamond**

Name ______________________ Date ______________ Class ____________

CHAPTER 20 REVIEW

Carbon and Hydrocarbons

SECTION 20-2

SHORT ANSWER **Answer the following questions in the space provided.**

1. Explain why the following two molecules are *not* isomers of one another.

```
     H   Cl                  H   H
     |   |                   |   |
 H — C — C — H           H — C — C — H
     |   |                   |   |
     Cl  H                   Cl  Cl
```

Free rotation around the single bond between the carbon atoms will allow both of these configurations to occur with the same molecule.

2. **a.** In the space below, draw the structural formula for two structural isomers with the same molecular formula.

Answers will vary.

```
                                             H
                                             |
     H   H   H   H                   H   H — C — H   H
     |   |   |   |                   |       |       |
 H — C — C — C — C — H           H — C ————— C ————— C — H
     |   |   |   |                   |       |       |
     H   H   H   H                   H       H       H
```

b. In the space below, draw the structural formula for two geometric isomers with the same molecular formula.

Answers will vary.

```
 H         Cl            Cl         Cl
   \      /                \       /
    C == C                  C == C
   /      \                /      \
 Cl        H             H         H
```

SECTION 20-2 continued

3. Draw a structural formula that demonstrates the catenation of the methane molecule, CH_4.

Answers will vary: any long hydrocarbon will do.

```
  H   H   H   H   H   H
  |   |   |   |   |   |
H—C — C — C — C — C — C—H
  |   |   |   |   |   |
  H   H   H   H   H   H
```

4. Draw the structural formula for two structural isomers of C_4H_{10}.

```
  H   H   H   H              H   H   H
  |   |   |   |              |   |   |
H—C — C — C — C—H          H—C — C — C—H
  |   |   |   |              |   |   |
  H   H   H   H              H   |   H
                               H—C—H
                                 |
                                 H
```

5. Draw the structural formula for the *cis*-isomer of $C_2H_2Cl_2$.

```
Cl       Cl
  \     /
   C = C
  /     \
H         H
```

6. Draw the structural formula for the *trans*-isomer of $C_2H_2Cl_2$.

```
Cl        H
  \     /
   C = C
  /     \
H        Cl
```

Name ______________________ Date ______________ Class ______________

CHAPTER 20 REVIEW

Carbon and Hydrocarbons

SECTION 20-3

SHORT ANSWER **Answer the following questions in the space provided.**

1. What is a saturated hydrocarbon?

a hydrocarbon in which each carbon atom forms four single covalent bonds with other atoms

2. Explain why the general formula for an alkane, C_nH_{2n+2}, correctly predicts hydrocarbons in a homologous series.

Each nonterminal carbon atom within the hydrocarbon chain bonds with two hydrogen atoms. The two terminal carbon atoms on the chain bond with an additional hydrogen atom each to complete carbon's four covalent bonds.

3. Why is the general formula for cyclic alkanes, C_nH_{2n}, different from the general formula for straight-chain hydrocarbons?

There are no terminal end carbons requiring a third hydrogen in a cyclic alkane.

4. Write the IUPAC name for the following structural formulas:

propane **a.**

```
   H   H   H
   |   |   |
H—C—C—C—H
   |   |   |
   H   H   H
```

3-methylpentane **b.**

```
   H   H       H       H   H
   |   |       |       |   |
H—C—C———C———C—C—H
   |   |       |       |   |
   H   H   H—C—H   H   H
               |
               H
```

3,4-diethylhexane **c.**

$$\begin{array}{c} CH_3—CH_2 \quad CH_2—CH_3 \\ | \qquad\quad | \\ CH_3—CH_2—CH—CH—CH_2—CH_3 \end{array}$$

SECTION 20-3 continued

3-ethyl-3-methylpentane **d.**

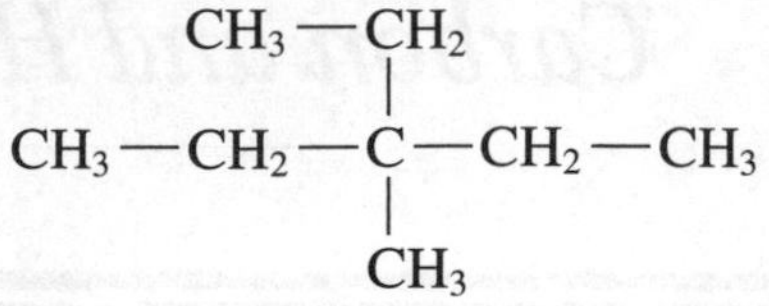

```
             CH3—CH2
                  |
    CH3—CH2—C—CH2—CH3
                  |
                 CH3
```

2-methylbutane **e.**

```
CH3—CH—CH2—CH3
       |
      CH3
```

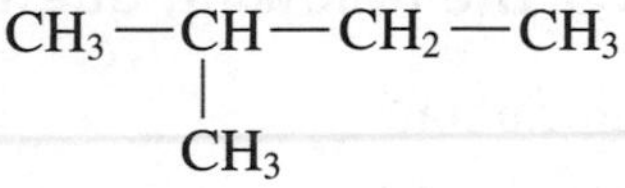

5. Draw the structural formula for each of the following compounds:

a. 2,3-dimethylbutane

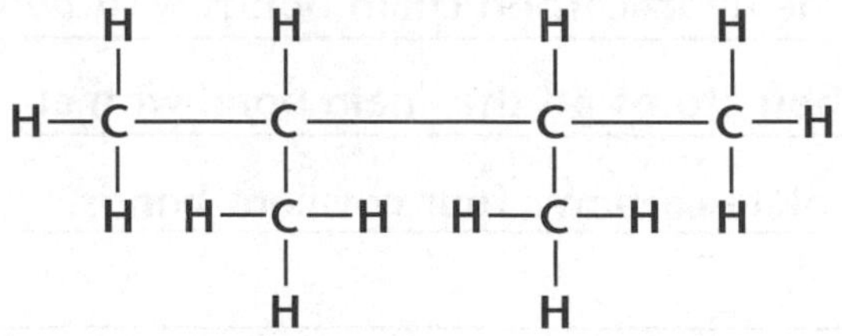

```
   H      H          H      H
   |      |          |      |
H—C——————C——————————C——————C—H
   |      |          |      |
   H   H—C—H      H—C—H     H
          |          |
          H          H
```

b. 2,2-dimethylpropane

```
          H
          |
   H   H—C—H   H
   |      |     |
H—C——————C—————C—H
   |      |     |
   H   H—C—H   H
          |
          H
```

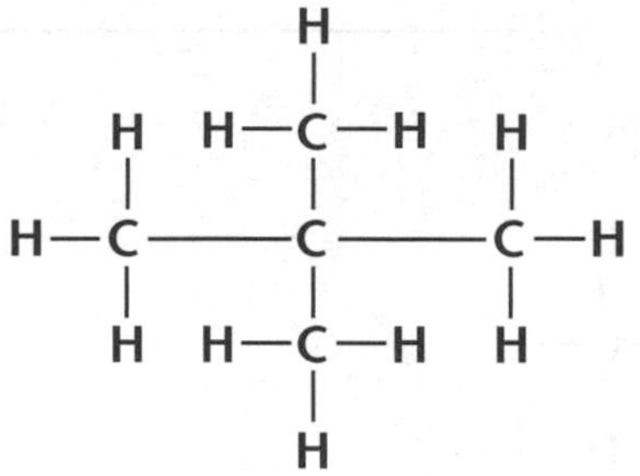

c. 3,4-diethyl-2-methylhexane

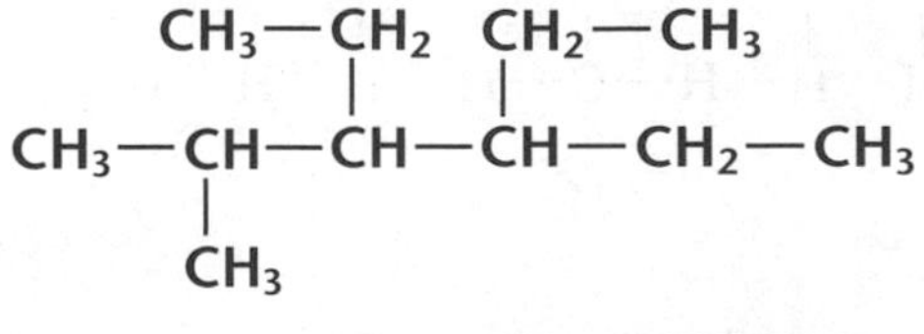

```
        CH3—CH2  CH2—CH3
               |    |
CH3—CH—CH—CH—CH2—CH3
       |
      CH3
```

Name ______________________ Date ______________ Class ______________

CHAPTER 20 REVIEW

Carbon and Hydrocarbons

SECTION 20-4

SHORT ANSWER **Answer the following questions in the space provided.**

1. b Hydrocarbons that contain double bonds are referred to as ____.

(a) alkanes
(b) alkenes
(c) alkynes
(d) aromatic

2. c Hydrocarbons that contain triple bonds are referred to as ____.

(a) alkanes
(b) alkenes
(c) alkynes
(d) aromatic

3. a Alkenes, alkynes, and aromatic hydrocarbons are all considered to be ____.

(a) unsaturated
(b) saturated
(c) homologous
(d) polar

4. b Alkenes and alkynes do not dissolve in water and are ____.

(a) polar
(b) nonpolar
(c) gases
(d) aromatic

5. What is an unsaturated hydrocarbon?

a hydrocarbon in which 2 or more carbon atoms in the chain are linked by double or triple bonds

6. Why are aromatic hydrocarbons less reactive than alkenes and alkynes?

Aromatic hydrocarbons are stabilized by resonance of delocalized electrons.

7. Write the IUPAC name for the following structural formulas:

1,2-dimethylbenzene **a.**

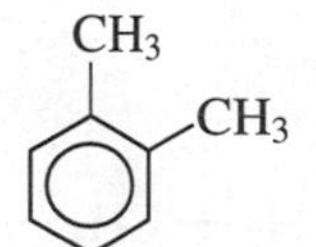

1-butyl-3-methylbenzene **b.**

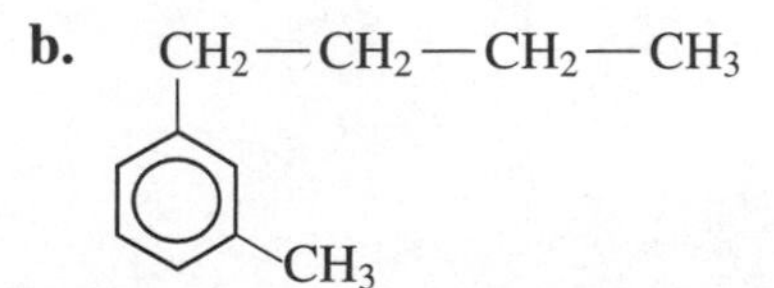

SECTION 20-4 continued

cis-2-butene ______________ **c.** 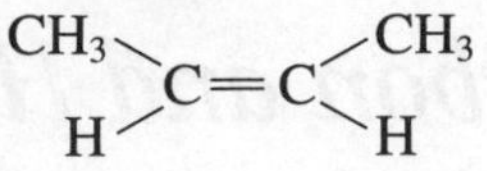 CH_3 / H — C=C — CH_3 / H (cis)

$$\begin{matrix} CH_3 & & CH_3 \\ & C=C & \\ H & & H \end{matrix}$$

trans-2-pentene ______________ **d.**

$$\begin{matrix} CH_3 & & H \\ & C=C & \\ H & & CH_2-CH_3 \end{matrix}$$

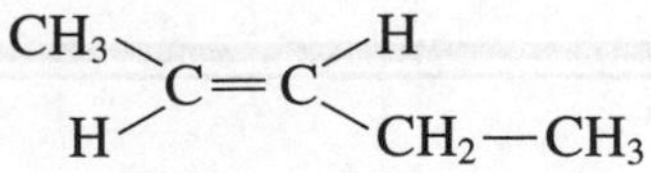

8. Draw the structural formula for each of the following compounds:

a. 1-ethyl-2,3-dimethylbenzene

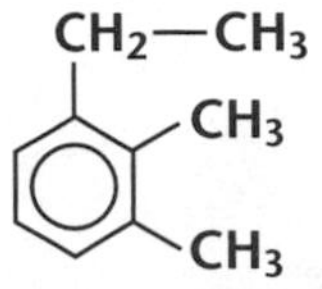

Benzene ring with CH_2-CH_3, CH_3, CH_3 substituents on adjacent carbons

b. *trans*-3-hexene

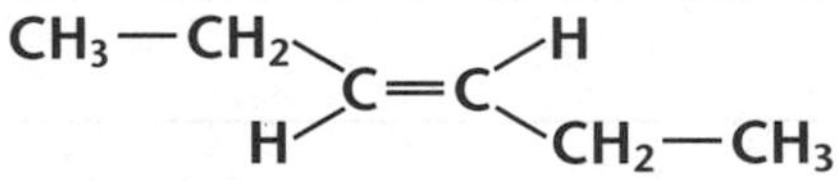

$$\begin{matrix} CH_3-CH_2 & & H \\ & C=C & \\ H & & CH_2-CH_3 \end{matrix}$$

c. *cis*-2-heptene

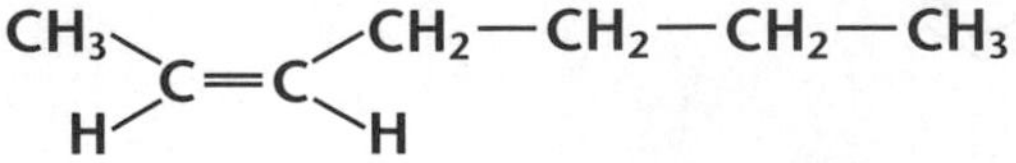

$$\begin{matrix} CH_3 & & CH_2-CH_2-CH_2-CH_3 \\ & C=C & \\ H & & H \end{matrix}$$

Name ______________________ Date ______________ Class ______________

CHAPTER 20 REVIEW

Carbon and Hydrocarbons

MIXED REVIEW

SHORT ANSWER **Answer the following questions in the space provided.**

1. Identify the hybridization of carbon in each of the following molecules:

sp^3 **a.** C_2H_6

sp **b.** HCN

sp^2 **c.** CH_2O

2. Arrange the following in order of increasing boiling point:

1 **a.** ethane

3 **b.** pentane

2 **c.** 2, 2-dimethylpropane

3. Explain why hydrocarbons with only single bonds cannot form geometric isomers.

The free rotation around the single bonds between carbon atoms prevents molecules with the same sequence of atoms from having different orientations in space.

4. How does increasing the length of a hydrocarbon chain affect the boiling point?

Increasing the number of carbons on a hydrocarbon chain increases the molecular mass, which increases the London dispersion forces necessary to hold the molecule together. Consequently, more energy is required to pull the molecule apart, raising the boiling point.

5. Write the IUPAC name for the following structural formulas:

a. **4-ethyl-3,5-dimethylheptane**

```
CH3—CH2  CH2—CH3
     |    |
CH3—CH—CH—CH—CH2—CH3
             |
            CH3
```

b. **1-hexene** $CH_2{=}CH{-}CH_2{-}CH_2{-}CH_2{-}CH_3$

Name ______________________ Date ______________ Class ______________

MIXED REVIEW continued

c. **3-methyl-1-butene**

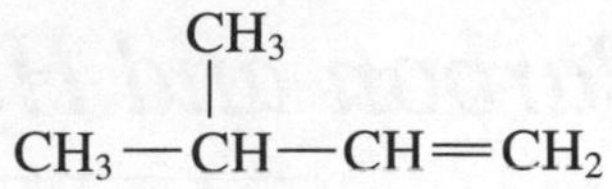

d. **2-butyne** $CH_3—C{\equiv}C—CH_3$

6. Draw the structural formula for each of the following compounds:

a. 1-ethyl-3-methylbenzene

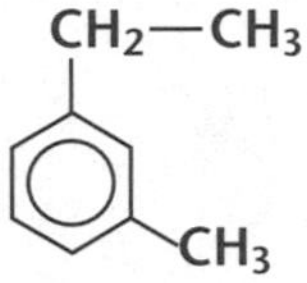

b. 1,2,4-trimethylcyclohexane

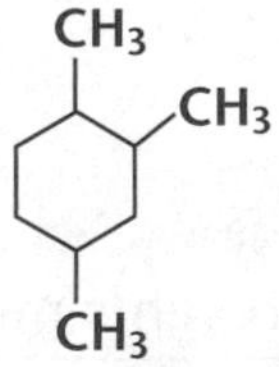

c. 3-methyl-1-pentyne

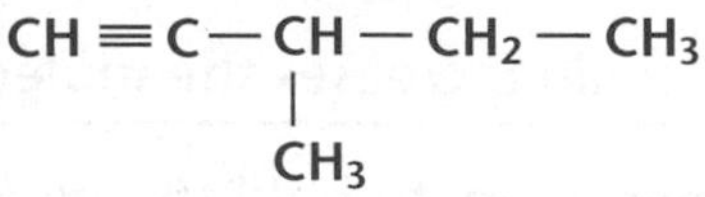

d. 3-ethyl-2-methylhexane

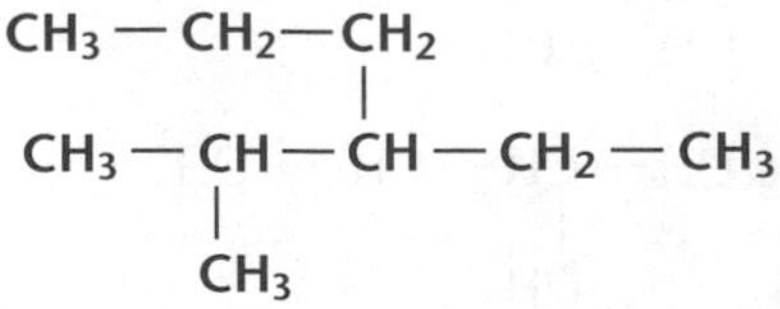

Name ______________________ Date ______________ Class ______________

CHAPTER 21 REVIEW

Other Organic Compounds

SECTION 21-1

SHORT ANSWER **Answer the following questions in the space provided.**

1. Draw the structural formula for each of the following compounds:

a. 1-butanol

```
    H   H   H   H
    |   |   |   |
H—C—C—C—C—H
    |   |   |   |
    H   H   H   OH
```

b. 1,1,2-trifluoroethene

```
F        F
 \      /
  C==C
 /      \
F        H
```

c. dipropyl ether

```
    H   H   H       H   H   H
    |   |   |       |   |   |
H—C—C—C—O—C—C—C—H
    |   |   |       |   |   |
    H   H   H       H   H   H
```

2. Write the IUPAC name for the following structural formulas:

2-propanol **a.**

```
    H   H   H
    |   |   |
H—C—C—C—H
    |   |   |
    H   OH  H
```

SECTION 21-1 continued

1,2-propanediol ______ **b.**

```
     H   H   H
     |   |   |
  H—C—C—C—H
     |   |   |
    OH  OH  H
```

3. There are four isomers in the alcohol family that fit the formula C_4H_9OH. Draw the structural formula for each isomer, and give each its corresponding IUPAC name.

1-butanol

```
     H   H   H   H
     |   |   |   |
  H—C—C—C—C—OH
     |   |   |   |
     H   H   H   H
```

2-butanol

```
     H   H   H   H
     |   |   |   |
  H—C—C—C—C—H
     |   |   |   |
     H   H  OH  H
```

2-methyl-2-propanol

```
     H    OH    H
     |    |     |
  H—C————C————C—H
     |    |     |
     H    |     H
          |
       H—C—H
          |
          H
```

2-methyl-1-propanol

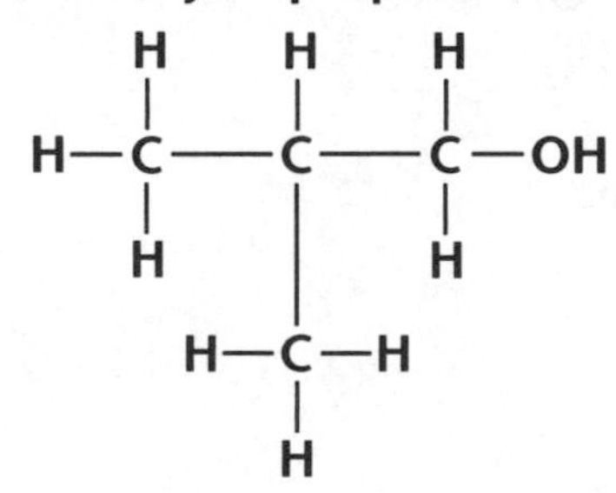

PROBLEM Write the answer on the line to the left. Show all your work in the space provided.

4. MTBE, an important gasoline additive, is discussed in Section 21-1 of the text. Its formula is $(CH_3)_3COCH_3$.

88.17 g/mol ______ **a.** What is its molar mass?

22 g ______ **b.** What is the mass of 0.25 mol of MTBE?

Name ______________________ Date ______________ Class ______________

CHAPTER 21 REVIEW

Other Organic Compounds

SECTION 21-2

SHORT ANSWER **Answer the following questions in the space provided.**

1. Match the structural formula on the right to the corresponding family name on the left.

__d__ aldehyde

__f__ ketone

__a__ carboxylic acid

__b__ amine

__c__ ester

__e__ alkene

(a)
```
H—C—OH
  ‖
  O
```

(b)
```
   H   H   H
   |   |   |
H—C—N—C—H
   |       |
   H       H
```

(c)
```
            H
            |
H—C—O—C—H
   ‖        |
   O        H
```

(d)
```
    H
    |
H—C—C=O
    |   |
    H   H
```

(e)
```
H       H
  \   /
   C=C
  /   \
H       H
```

(f)
```
    H  O  H
    |  ‖  |
H—C—C—C—H
    |      |
    H      H
```

2. Draw the structural formula for each of the following compounds:

a. 2-pentanone

```
   H  O  H  H  H
   |  ‖  |  |  |
H—C—C—C—C—C—H
   |     |  |  |
   H     H  H  H
```

b. 3-methyl hexanoic acid

```
   H  H  H     H     H
   |  |  |     |     |
H—C—C—C——C——C—C=O
   |  |  |     |     |  |
   H  H  H H—C—H  H  OH
               |
               H
```

SECTION 21-2 continued

c. propyl amine

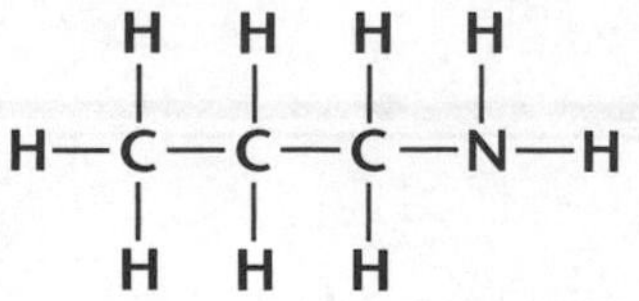

```
    H   H   H   H
    |   |   |   |
H — C — C — C — N — H
    |   |   |
    H   H   H
```

d. methyl butanoate

```
    H   H   H   O       H
    |   |   |   ‖       |
H — C — C — C — C — O — C — H
    |   |   |           |
    H   H   H           H
```

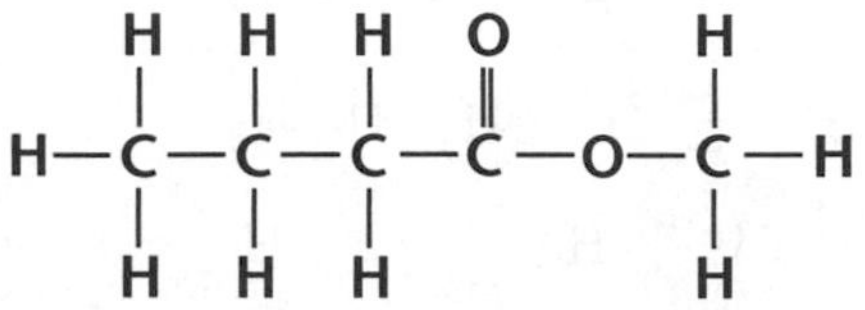

3. Write the IUPAC name for the following structural formulas:

propanal **a.**

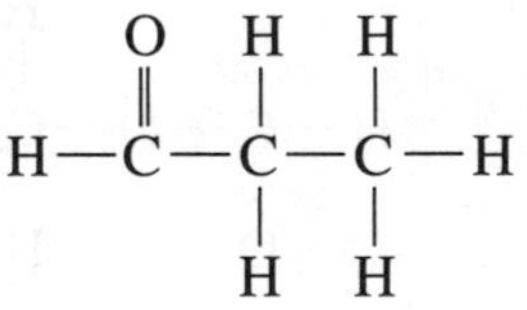

```
    O   H   H
    ‖   |   |
H — C — C — C — H
        |   |
        H   H
```

ethanoic acid **b.**

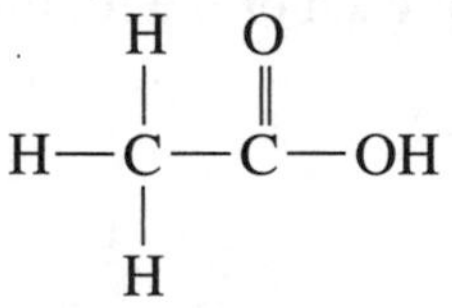

```
    H   O
    |   ‖
H — C — C — OH
    |
    H
```

3-bromo-2-pentanone **c.**

```
    H   O   Br  H   H
    |   ‖   |   |   |
H — C — C — C — C — C — H
    |       |   |   |
    H       H   H   H
```

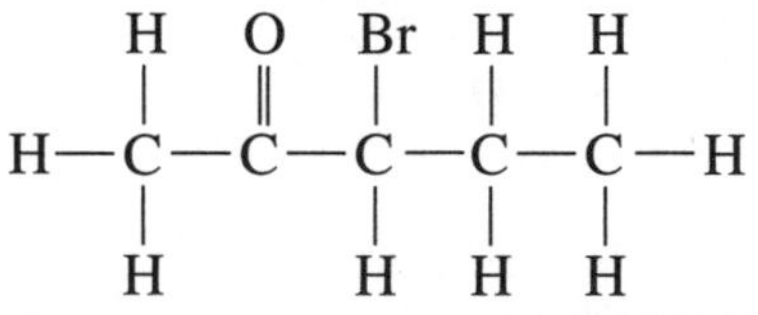

2-methyl-1-propanamine **d.**

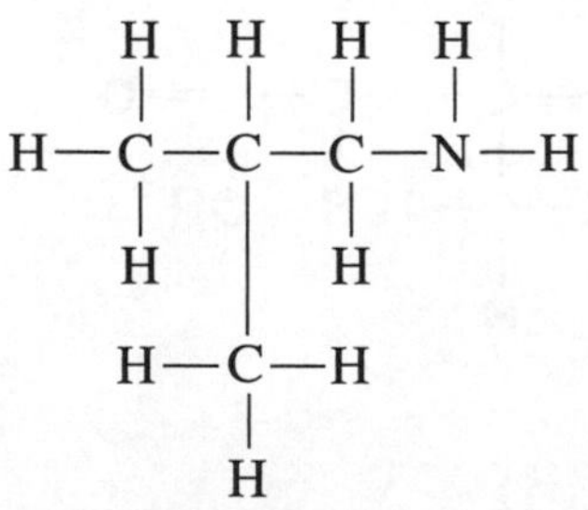

```
    H   H   H   H
    |   |   |   |
H — C — C — C — N — H
    |   |   |
    H   |   H
        |
    H — C — H
        |
        H
```

Name ______________________ Date ______________ Class ______________

CHAPTER 21 REVIEW

Other Organic Compounds

SECTION 21-3

SHORT ANSWER **Answer the following questions in the space provided.**

1. ___a___ A saturated organic compound ____.

(a) contains all single bonds
(b) contains at least one double or triple bond
(c) contains only carbon and hydrogen atoms
(d) is quite soluble in water

2. Match the reaction type on the left to its description on the right.

___c___ substitution **(a)** An atom or molecule is added to an unsaturated molecule, increasing the saturation of the molecule.

___a___ addition **(b)** A simple molecule is removed from adjacent carbon atoms of a larger molecule.

___d___ condensation **(c)** One or more atoms replace another atom or group of atoms in a molecule.

___b___ elimination **(d)** Two molecules or parts of the same molecule combine.

3. Substitution reactions can require a catalyst to be feasible. The reaction represented by the following equation is heated to maximize the percent yield:

$$C_2H_6(g) + Cl_2(g) \overset{\Delta}{\rightleftarrows} C_2H_5Cl(l) + HCl(g)$$

___high___ **a.** Should a high or low temperature be maintained?

___high___ **b.** Should a high or low pressure be used?

___Yes___ **c.** Should the HCl gas be allowed to escape?

4. Elemental bromine is a red-brown liquid that is colorless when it exists in compounds. A qualitative test for carbon-carbon multiple bonds is to add a few drops of bromine solution to a hydrocarbon sample at room termperature and in the absence of sunlight. The bromine will either quickly lose its color or remain reddish brown.

___addition___ **a.** If the sample is unsaturated, what type of reaction should occur when the bromine is added under the conditions mentioned above?

___no reaction___ **b.** If the sample is saturated, what type of reaction should occur when the bromine is added under the conditions mentioned above?

___unsaturated___ **c.** The red-brown color of a bromine solution added to a hydrocarbon sample at room temperature and in the absence of light quickly disappears. Is the sample a saturated or unsaturated hydrocarbon?

SECTION 21-3 continued

5. Write the formulas for the products of the following reactions:

a. addition of chlorine to ethene

$C_2H_4Cl_2$

b. substitution of iodine with ethane

$C_2H_5I + HI$

c. elimination of water from

```
   H   H
   |   |
H—C—C—H
   |   |
   H   OH
```

$C_2H_4 + H_2O$

6. Two glucose molecules, $C_6H_{12}O_6$, undergo a condensation reaction to form one molecule of sucrose, $C_{12}H_{22}O_{11}$.

1 **a.** How many molecules of water split off during this condensation reaction?

b. Write a balanced formula equation for this condensation reaction.

$2C_6H_{12}O_6 \rightarrow C_{12}H_{22}O_{11} + H_2O$

7. Addition reactions with halogens tend to proceed rapidly and easily with the two halogen atoms bonding to the carbon atoms connected by the multiple bond. Thus, only one isomeric product forms.

a. Write an equation showing the structural formulas for the reaction of Br_2 with 1-butene.

```
   H  H  H  H                H  H  H  H
   |  |  |  |                |  |  |  |
H—C=C—C—C—H + Br2→H—C—C—C—C—H
         |  |                |  |  |  |
         H  H               Br Br  H  H
```

b. Name the product.

1,2-dibromobutane

1:1 **c.** What is the mole ratio of 1-butene to Br_2 in the above reaction?

d. How many moles of Br_2 will add to 1 mol of 1-butyne? Explain your answer.

2 mol; The first mole of Br_2 converts the triple bond of the alkyne to a double bond, and the second mole of Br_2 converts the double bond of the alkene to a single bond.

Name ______ Date ______ Class ______

CHAPTER 21 REVIEW

Other Organic Compounds

SECTION 21-4

SHORT ANSWER **Answer the following questions in the space provided.**

1. Identify each of the following substances as either a natural or a synthetic polymer:

natural a. cellulose

synthetic b. nylon

natural c. proteins

natural d. polyisoprene

synthetic e. kevlar

synthetic f. polypropylene

2. Table 21-8 on page 687 of the text shows the styrene monomer, $CH_2{=}CH{-}R$, where *R* = (benzene ring).

C_6H_5 a. What is the formula of *R* in this molecule?

104.16 g/mol b. What is the molar mass of this styrene monomer?

3. The text gives several abbreviations commonly used in describing plastics or polymers. For each of the following abbreviations, give the complete name and one common usage:

a. HDPE

high-density polyethylene; rigid plastic bottles

b. LDPE

low-density polyethylene; plastic shopping bags

c. CLPE

cross-linked polyethylene; plastic caps on soda bottles

d. PVA

polyvinyl acetate; latex paints

e. PVC

polyvinyl chloride; plastic pipes for plumbing

f. SBR

styrene-butadiene rubber; tires

Name ______________________ Date ______________ Class ______________

SECTION 21-4 continued

g. PET

polyethylene terephthalate; food packaging

4. Sulfur is often used in making cross-linked polymers (vulcanized rubber, for example).

a. Give the orbital electron arrangement of a neutral sulfur atom.

$1s^2 2s^2 2p^6 3s^2 3p^4$ or $[Ne]3s^2 3p^4$

2 b. How many unpaired electrons are there in this atom?

c. Explain how sulfur links two adjacent chains of polyethylene.

Sulfur atoms bond to a carbon atom in one molecule and a second carbon atom in a second molecule, forming a cross-link between the two molecular chains.

5. Based on the description of natural rubber in the text, is natural rubber a thermoplastic or thermosetting polymer? Explain your answer.

Natural rubber is a thermoplastic polymer because it melts when heated.

6. Explain why an alkane cannot be used as the monomer of an addition polymer.

There must be a multiple bond onto which an adjoining CH_2 group can add. An alkane has all single bonds.

Name ______________________ Date ______________ Class ______________

CHAPTER 21 REVIEW
Other Organic Compounds

MIXED REVIEW

SHORT ANSWER **Answer the following questions in the space provided.**

1. Match the general formula on the right to the corresponding family name on the left.

e carboxylic acid

h ester

d alcohol

c ether

a alkyl halide

f amine

b aldehyde

g ketone

(a) *R*—X

(b) *R*—C(=O)—H

(c) *R*—O—*R*′

(d) *R*—OH

(e) *R*—C(=O)—OH

(f) *R*—N(—*R*′)—*R*″

(g) *R*—C(=O)—*R*′

(h) *R*—C(=O)—O—*R*′

2. Write the IUPAC name for the following structural formulas:

pentanal **a.**
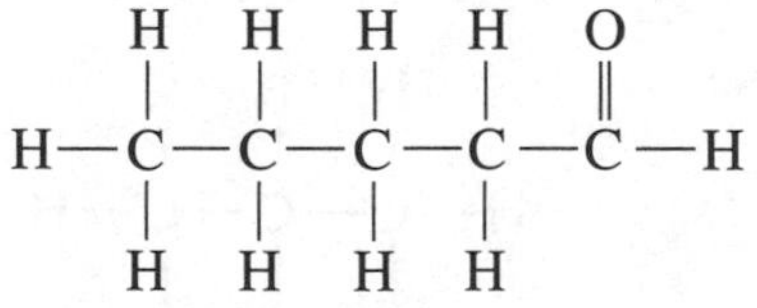

propanoic acid **b.**
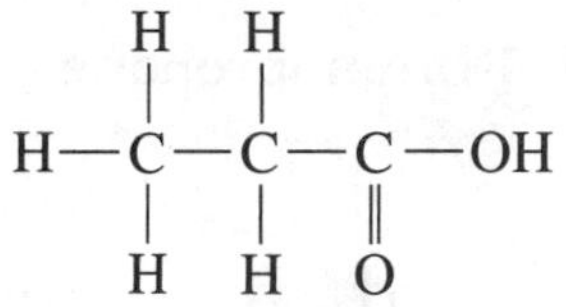

1,2,4-trichlorobutane **c.**
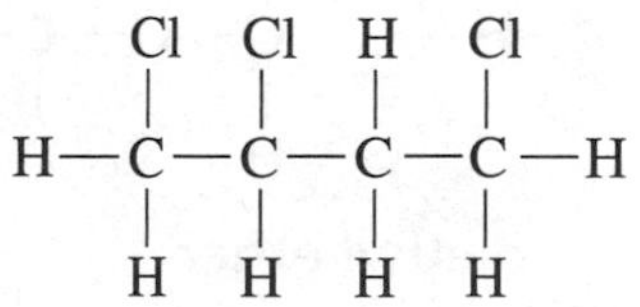

2-methyl-2-propanol **d.**
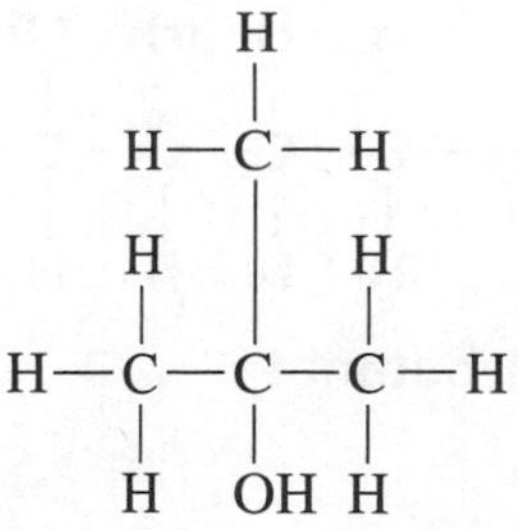

MIXED REVIEW continued

3. Write the name of the organic product formed by the following reactions:

2,3-dibromobutane ______ **a.** addition of bromine to 2-butene

propene ______ **b.** elimination of water from 1-propanol

fluoromethane ______ **c.** reaction of methane and F_2 in direct sunlight

4. Recall that isomers in organic chemistry have identical molecular formulas but different structures and IUPAC names.

True ______ **a.** Two isomers must have the same molar mass. True or False?

False ______ **b.** Two isomers must have the same boiling point. True or False?

Yes ______ **c.** Are ethanol and dimethyl ether isomers? Support your answer by drawing the structural formula for each compound and labeling it.

```
    H  H                    H        H
    |  |                    |        |
 H—C—C—OH                H—C—O—C—H
    |  |                    |        |
    H  H                    H        H
   ethanol              dimethyl ether
```

5. Each of the following names implies a structure but is not a correct IUPAC name. For each example, draw the implied structural formula and write the correct IUPAC name.

a. 3-bromopropane

```
     H  H  H
     |  |  |
  H—C—C—C—H
     |  |  |
    Br  H  H
```

1-bromopropane

b. ethylethylether

```
     H  H     H  H
     |  |     |  |
  H—C—C—O—C—C—H
     |  |     |  |
     H  H     H  H
```

diethyl ether

c. 4-butanol

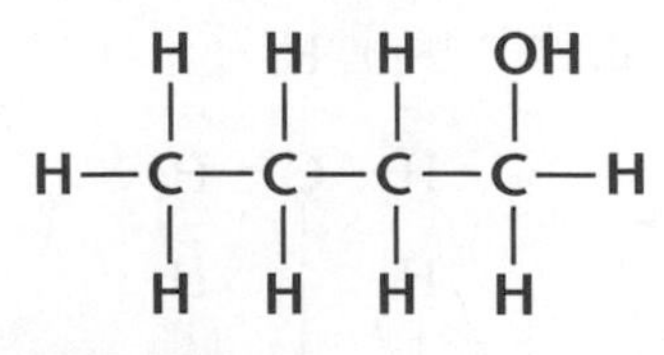

1-butanol

Name ______________________ Date ______________ Class ______________

CHAPTER 22 REVIEW
Nuclear Chemistry

SECTION 22-1

SHORT ANSWER **Answer the following questions in the space provided.**

1. __b__ Based on the masses of the three elementary particles reported in Section 22-1 on page 701 of the text, which has the greatest mass?

(a) the proton
(b) the neutron
(c) the electron

2. __a__ The force that keeps nucleons together is ____.

(a) a strong force
(b) a weak force
(c) an electromagnetic force
(d) gravity

3. __d__ The stability of a nucleus is most affected by the ____.

(a) number of neutrons
(b) number of protons
(c) number of electrons
(d) ratio of neutrons to protons

4. __b__ If an atom should form from its constituent particles, ____.

(a) matter is lost and energy is taken in
(b) matter is lost and energy is released
(c) matter is gained and energy is taken in
(d) matter is gained and energy is released

5. __b__ For atoms of a given mass number, the greater the mass defect, the ____.

(a) smaller the binding energy per nucleon
(b) greater the binding energy per nucleon
(c) binding energy per nucleon stays the same

6. Based on the data in Table 22-1 on page 705 of the text, which isotope of He, helium-3 or helium-4,

__helium-3__ **a.** has the smaller binding energy per nucleon?

__helium-4__ **b.** is more stable to nuclear changes?

7. __13__ The number of neutrons in an atom of magnesium-25 is ____.

8. __$(A - Z)$__ The mass number of a given nuclide is *A*, and its atomic number is *Z*. Use these two variables to write the formula to calculate the number of neutrons in a nuclide.

SECTION 22-1 continued

9. Atom X has 50 nucleons and a binding energy of 4.2×10^{-11} J. Atom Z has 80 nucleons and a binding energy of 8.4×10^{-11} J.

True **a.** The mass defect of Z is twice that of X. True or False?

atom Z **b.** Which atom has the greater binding energy per nucleon?

atom Z **c.** Which atom is likely to be more stable to nuclear transmutations?

10. Identify the missing term in the following nuclear equations. Write the element's symbol, its atomic number, and its mass number.

$^{14}_{7}N$ **a.** $^{14}_{6}C \rightarrow {}^{0}_{-1}e + \_\_\_\_$

$^{60}_{28}Ni$ **b.** $^{63}_{29}Cu + {}^{1}_{1}H \rightarrow \_\_\_\_ + {}^{4}_{2}He$

11. Write the equation that shows the equivalency of mass and energy.

$E = mc^2$

12. Consider the two nuclides $^{56}_{26}Fe$ and $^{14}_{6}C$.

a. Determine the number of protons in each nuclei.

Iron-56 has 26 protons; carbon-14 has 6 protons.

b. Determine the number of neutrons in each nuclei.

Iron-56 has 30 neutrons; carbon-14 has 8 neutrons.

c. Determine whether the $^{56}_{26}Fe$ nuclide is likely to be stable or unstable, based on its position in the band of stability shown in Figure 22-2 on page 703 of the text.

Iron-56 is likely to be stable.

PROBLEM Write the answer on the line to the left. Show all your work in the space provided.

13. 0.172 46 amu Neon-20 is a stable isotope of neon. Its actual mass has been found to be 19.992 44 amu. Determine the mass defect in this nuclide.

Name ________________________ Date ______________ Class ______________

CHAPTER 22 REVIEW

Nuclear Chemistry

SECTION 22-2

SHORT ANSWER **Answer the following questions in the space provided.**

1. __a__ The nuclear equation $^{210}_{84}Po \rightarrow ^{206}_{82}Pb + ^{4}_{2}He$ is an example of ____.

(a) alpha emission
(b) beta emission
(c) positron emission
(d) electron capture

2. __d__ $^{a}_{b}Z$ undergoes electron capture to form a new element X. Which of the following choices best represents the product formed?

(a) $^{a-1}_{b}X$
(b) $^{a+1}_{b}X$
(c) $^{a}_{b+1}X$
(d) $^{a}_{b-1}X$

3. __a__ Which of the following choices best represents the fraction of a radioactive sample remaining after four half-lives have occurred?

(a) $(1/2)^4$
(b) $(1/2) \times 4$
(c) $(1/4)$
(d) $(1/4)^2$

4. Match the nuclear symbol on the right to the name of the corresponding particle on the left.

__c__ beta particle **(a)** $^{1}_{1}H$

__a__ proton **(b)** $^{4}_{2}He$

__d__ positron **(c)** $^{0}_{-1}\beta$

__b__ alpha particle **(d)** $^{0}_{+1}\beta$

5. Label each of the following statements as True or False. Each member of a decay series ____.

__True__ **a.** shares the same atomic number

__True__ **b.** differs in mass number from others by multiples of 4

__False__ **c.** has a unique atomic number

__False__ **d.** differs in atomic number from others by multiples of 4

SECTION 22-2 continued

6. $^{235}_{92}U$ Identify the missing term in the following nuclear equation. Write the element's symbol, its atomic number, and its mass number.

$$\_\_\_\_ \rightarrow {}^{231}_{90}Th + {}^{4}_{2}He$$

7. Lead-210 undergoes beta emission. Write the nuclear equation showing this transmutation.

$$^{210}_{82}Pb \rightarrow {}^{0}_{-1}\beta + {}^{210}_{83}Bi$$

8. Einsteinium is a transuranium element. Einsteinium-247 can be prepared by bombarding uranium-238 with nitrogen-14 nuclei, releasing several neutrons, as shown by the following equation:

$$^{238}_{92}U + {}^{14}_{7}N \rightarrow {}^{247}_{99}Es + x\,{}^{1}_{0}n$$

What must be the value of x in the above equation? Explain your reasoning.

x = 5; The total mass of the reactants is 252 amu, therefore the total mass of the products must be 252. (252 − 247) = 5 amus left for neutrons at 1 amu each

PROBLEMS Write the answer on the line to the left. Show all your work in the space provided.

9. **42.9 days** Phosphorus-32 has a half-life of 14.3 days. How many days will it take for a radioactive sample to decay to one-eighth its original size?

10. **5.0 mg** Iodine-131 has a half-life of 8.0 days. How many grams of an original 160 mg sample will remain in 40 days?

11. **0.24 mg** Carbon-14 has a half-life of 5715 years. It is used to determine the age of ancient objects. If a sample today contains 0.060 mg of carbon-14, how much carbon-14 must have been present in the sample 11 430 years ago?

Name ______________________ Date ______________ Class ______________

CHAPTER 22 REVIEW

Nuclear Chemistry

SECTION 22-3

SHORT ANSWER **Answer the following questions in the space provided.**

1. d The radioisotope cobalt-60 is used for all of the following applications *except* to ____.

(a) kill food-spoiling bacteria
(b) preserve food
(c) kill insects that infest food
(d) treat heart disease
(e) treat certain kinds of cancers

2. c All of the following contribute to background-radiation exposure *except* ____.

(a) radon in homes and buildings
(b) cosmic rays passing through the atmosphere
(c) consumption of irradiated foods
(d) X rays obtained for medical or dental reasons
(e) rocks in Earth's soil

3. b Review the Dating Game Commentary on page 716 of the text. Which of the graphs shown below best illustrates the decay of a sample of carbon-14? Assume each division on the time axis represents 5715 years.

(a)

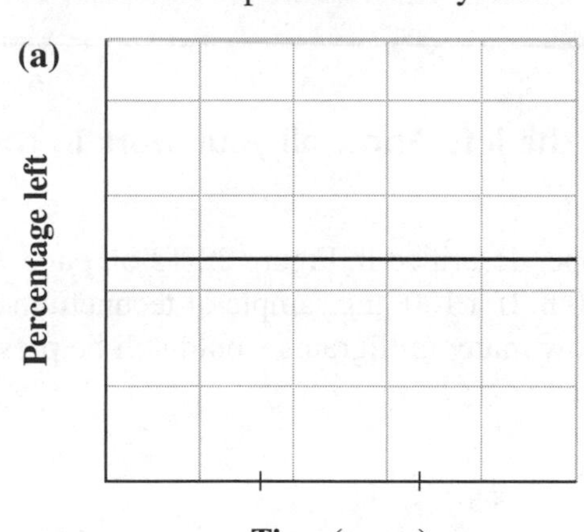

(c)

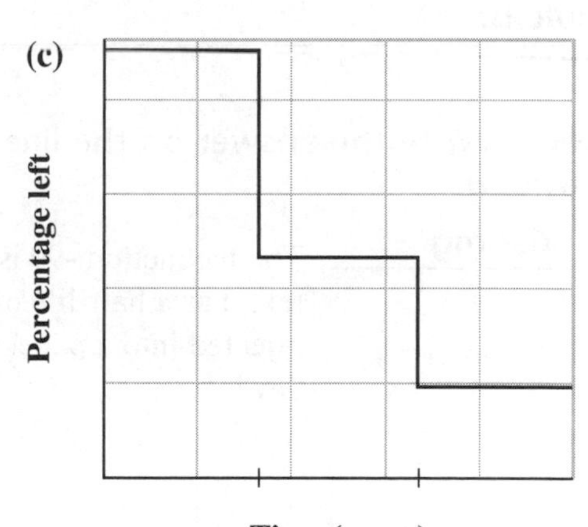

(b)

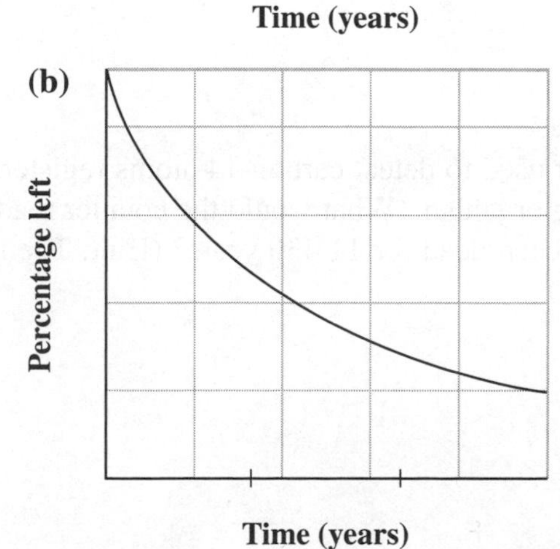

(d)

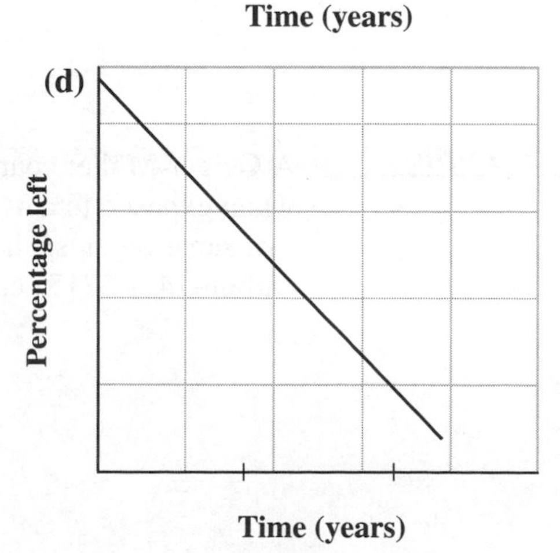

4. Match the item on the left with its description on the right.

<u>c</u> Geiger-Müller counter

<u>b</u> scintillation counter

<u>a</u> film badge

<u>d</u> radioactive tracer

(a) uses exposure to film to measure the approximate radiation exposure of people working with radiation

(b) instrument that converts scintillating light to an electric signal for detecting radiation

(c) detects radiation by counting electric pulses carried by gas ionized by radiation

(d) radioactive atoms that are incorporated into substances so that movement of the substances can be followed by detectors

5. Which type of radiation is most easily absorbed by shielding? Why?

Alpha radiation penetrates the least; therefore, it is most easily absorbed.

6. One technique for dating ancient rocks involves uranium-235, which has a half-life of 710 million years. Rocks originally rich in uranium-235 will contain small amounts of its decay series, including the nonradioactive lead-206. Explain the relationship between the ratio of lead-206 to uranium-235 in a sample and the sample's age.

The greater the ratio of lead-206 to uranium-235 in the sample, the older the rock sample is.

PROBLEMS Write the answer on the line to the left. Show all your work in the space provided.

7. <u>6.2 mg</u> The technetium-99 isotope, described in Figure 22-13 on page 715 of the text, has a half-life of 6.0 h. If a 100. mg sample of technetium-99 were injected into a patient, how many milligrams would still be present after one day?

8. <u>3.5 units</u> A Geiger-Müller counter used to detect carbon-14 atoms registers 14 units when exposed to a living organism. What would the counter read in units if that organism had been dead for 11 430 years? (Hint: The half-life of carbon-14 is 5715 years.)

Name ______________________ Date ______________ Class ______________

CHAPTER 22 REVIEW

Nuclear Chemistry

SECTION 22-4

SHORT ANSWER **Answer the following questions in the space provided.**

1. Label each of the following statements with one of the choices below:

(1) fission only (3) both fission and fusion
(2) fusion only (4) neither fission nor fusion

__1__ **a.** A very large nucleus splits into smaller pieces.

__3__ **b.** The total mass before a reaction is greater than the mass after a reaction.

__3__ **c.** The rate of a reaction can be safely controlled in suitable vessels.

__2__ **d.** Two small nuclei form a single larger one.

__3__ **e.** Less-stable nuclei are converted to more-stable nuclei.

2. Match the reaction type on the right to the statement that applies to it on the left.

__c__	requires very high temperatures	**(a)** uncontrolled fusion
__d__	used in nuclear reactors to make electricity	**(b)** uncontrolled fission
__a__	used in hydrogen bombs	**(c)** controlled fusion
__b__	used in atomic bombs	**(d)** controlled fission

3. Match the component of a nuclear power plant on the right to its use on the left.

__c__	limits the number of free neutrons	**(a)** moderator
__a__	used to slow down neutrons	**(b)** fuel rod
__f__	drives an electrical generator	**(c)** control rod
__b__	provides neutrons by its fission	**(d)** shielding
__e__	removes heat from the system safely	**(e)** coolant
__d__	prevents escape of radiation	**(f)** turbine

4. __b__ A chain reaction is any reaction in which ____.

(a) excess reactant is present

(b) the material that starts the reaction is also a product

(c) the rate is slow

(d) many steps are involved

5. As a star ages, does the ratio of He atoms to H atoms in its composition become larger, smaller, or remain constant? Explain your answer.

The ratio of He atoms to H atoms in a star's composition becomes larger. As a star ages, H atoms fuse into He atoms. A new star is almost 100% H, but the older the star becomes, the more of its H is converted into He.

6. The products of nuclear fission are variable; many possible nuclides can be created. In the feature An Unexpected Finding, on page 720 of the text, it was noted that Meitner showed radioactive barium to be one product of fission. Following is an incomplete possible nuclear equation to produce barium-141:

$$^{235}_{92}\text{U} + ^{1}_{0}n \rightarrow ^{141}_{56}\text{Ba} + \_\_\_ + 4\,^{1}_{0}n + \text{energy}$$

$^{91}_{36}$Kr **a.** Determine the missing fission product formed. Write the element's symbol, its atomic number, and its mass number.

Yes **b.** Is it likely that this isotope in part a is unstable? (Refer to Figure 22-2 on page 703 of the text.)

7. Small nuclides other than hydrogen-1 can undergo fusion.

$^{10}_{4}$Be **a.** Complete the following nuclear equation by identifying the missing term. Write the element's symbol, its atomic number, and its mass number.

$$^{3}_{1}\text{H} + ^{7}_{3}\text{Li} \rightarrow \_\_\_ + \text{energy}$$

b. When measured exactly, the total mass of the reactants does not add up to that of the products in the reaction represented in part a. Why is there a difference between the mass of the products and the mass of the reactants? Which has the greater mass, the reactants or the products?

Some of the mass is converted into energy. The mass of the reactants is greater than the mass of the products.

8. What are some current concerns regarding nuclear-power-plant development?

Current concerns include environmental requirements, safety of the operation, plant construction costs, and storage of spent fuel.

Name ______________________ Date ______________ Class ______________

CHAPTER 22 REVIEW

Nuclear Chemistry

MIXED REVIEW

SHORT ANSWER **Answer the following questions in the space provided.**

1. The ancient alchemists dreamed of being able to turn lead into gold. By using lead-206 as the target atom of a powerful accelerator, we can attain that dream in principle. Find a one-step process that will convert $^{206}_{82}Pb$ into a nuclide of gold-79. You may use alpha particles, beta particles, positrons, or protons. Write the nuclear equation to turn lead into gold.

$^{206}_{82}Pb + ^{0}_{-1}e \rightarrow ^{4}_{2}He + ^{202}_{79}Au$

2. A typical fission reaction releases 2×10^{10} kJ/mol of uranium-235, while a typical fusion reaction produces 6×10^{8} kJ/mol of hydrogen-1. Which process produces more energy per gram of starting material? Explain your answer.

Fusion produces more energy per gram of starting material. 235 g of uranium-235 = 1.0 mol, and this releases 2×10^{10} J of energy in a fission reaction. 235 g of hydrogen-1 = 235 mol, and this releases $235(6 \times 10^{8}$ J) or 1.4×10^{11} J of energy in a fusion reaction, an increase of seven times.

3. Write the nuclear equations for the following reactions:

a. Carbon-12 combines with hydrogen-1 to form nitrogen-13.

$^{12}_{6}C + ^{1}_{1}H \rightarrow ^{13}_{7}N$

b. Curium-246 combines with carbon-12 to form nobelium-254 and four neutrons.

$^{246}_{96}Cm + ^{12}_{6}C \rightarrow ^{254}_{102}No + 4^{1}_{0}n$

c. Hydrogen-2 combines with hydrogen-3 to form helium-4 and a neutron.

$^{2}_{1}H + ^{3}_{1}H \rightarrow ^{4}_{2}He + ^{1}_{0}n$

4. Write the complete nuclear equations for the following reactions:

a. $^{91}_{42}Mo$ undergoes positron emission.

$^{91}_{42}Mo \rightarrow ^{91}_{41}Nb + ^{0}_{+1}\beta$

b. $^{6}_{2}He$ undergoes beta decay.

$^{6}_{2}He \rightarrow ^{6}_{3}Li + ^{0}_{-1}\beta$

c. $^{194}_{84}Po$ undergoes alpha decay.

$^{194}_{84}Po \rightarrow ^{190}_{82}Pb + ^{4}_{2}He$

MIXED REVIEW continued

d. $^{129}_{55}Cs$ undergoes electron capture.

$^{129}_{55}Cs + ^{0}_{-1}e \rightarrow ^{129}_{54}Xe$

PROBLEMS Write the answer on the line to the left. Show all your work in the space provided.

5. 2.2×10^{-12} J It was shown in Section 22-1 of the text that a mass defect of 0.030 38 amu corresponds to a binding energy of 4.54×10^{-12} J. What binding energy would a mass defect of 0.015 amu yield?

6. 24 days Iodine-131 has a half-life of 8.0 days; it is used in medical treatments for thyroid conditions. Determine how many days must elapse for a 0.80 mg sample of iodine-131 in the thyroid to decay to 0.10 mg.

7. Following is an incomplete nuclear fission equation:

$$^{235}_{92}U + ^{1}_{0}n \rightarrow ^{90}_{38}Sr + ^{141}_{54}Xe + x\,^{1}_{0}n + \text{energy}$$

5 **a.** Determine the value of x in the above equation.

1/8 **b.** The strontium-90 produced in the above reaction has a half-life of 28 years. What fraction of strontium-90 still remains in the environment 84 years after it is produced in the reactor?